BTEC National for IT Practitione
Core Units

WITHDRAWN

This material has been endorsed by Edexcel and offers high quality support for the delivery of Edexcel qualifications.

Edexcel endorsement does not mean that this material is essential to achieve any Edexcel qualification, nor does it mean that this is the only suitable material available to support any Edexcel qualification. No endorsed material will be used verbatim in setting any Edexcel examination and any resource lists produced by Edexcel shall include this and other appropriate texts. While this material has been through an Edexcel quality assurance process, all responsibility for the content remains with the publisher.

Copies of official specifications for all Edexcel qualifications may be found on the Edexcel website – www.edexcel.org.uk

BTEC National for IT Practitioners: Core Units

Common Core and Specialist Units for all Pathways

Sharon Yull

THE LIBRARY
NEW COLLEGE
SWINDON

ELSEVIER

AMSTERDAM • BOSTON • HEIDELBERG • LONDON • NEW YORK • OXFORD
PARIS • SAN DIEGO • SAN FRANCISCO • SINGAPORE • SYDNEY • TOKYO

Newnes is an imprint of Elsevier

Newnes

Newnes is an imprint of Elsevier

Linacre House, Jordan Hill, Oxford OX2 8DP, UK

30 Corporate Drive, Suite 400, Burlington, MA 01803, USA

First edition 2009

Copyright © 2009, Sharon Yull. Published by Elsevier Ltd. All rights reserved.

The right of Author Name to be identified as the author of this work has been
asserted in accordance with the Copyright, Designs and Patents Act 1988

No part of this publication may be reproduced, stored in a retrieval system or
transmitted in any form or by any means electronic, mechanical, photocopying,
recording or otherwise without the prior written permission of the publisher.

Permissions may be sought directly from Elsevier's Science & Technology Rights
Department in Oxford, UK: phone (+44) (0) 1865 843830; fax (+44) (0) 1865 853333;
email: permissions@elsevier.com. Alternatively you can submit your request online
by visiting the Elsevier web site at http://elsevier.com/locate/permissions, and selecting
Obtaining permission to use Elsevier material

Notice *2011000945*

No responsibility is assumed by the publisher for any injury and/or damage to
persons or property as a matter of products liability, negligence or otherwise, or from
any use or operation of any methods, products, instructions or ideas contained in the
material herein. Because of rapid advances in the medical sciences, in particular,
independent verification of diagnoses and drug dosages should be made

British Library Cataloguing in Publication Data
A catalogue record for this book is available from the British Library

Library of Congress Cataloging-in-Publications Data
A catalog record for this book is available from the Library of Congress

ISBN: 978-0-7506-8652-5

For information on all Newnes publications
visit our web site at books.elsevier.com

Typeset by Macmillan Publishing Solutions
(www.macmillansolutions.com)

Printed and bound in Slovenia
09 10 10 9 8 7 6 5 4 3 2 1

Working together to grow
libraries in developing countries

www.elsevier.com | www.bookaid.org | www.sabre.org

ELSEVIER   BOOK AID
International   Sabre Foundation

Dedication

I would like to dedicate this book to my daughter who is a constant source of inspiration.

Contents

Preface

Introduction

Welcome to the ever changing world of information and communications technology. This book has been designed to provide you with a range of knowledge, information and skills that will facilitate you in understanding the BTEC Nationals IT Practitioners qualification.

About the BTEC National Certificate and Diploma

The BTEC National Certificate and Diploma qualifications are Level 3 qualifications that have been designed to provide you with a range of practical skills and underpinning knowledge that will allow you to progress onto a higher level course or prepare you for a job in ICT and computing.

ICT is such a growing area that you will find all areas of the BTEC specification appropriate. The units have been designed with the support of practitioners, experts in the field and also in collaboration with industry. You will be able to use elements of the qualification in a range of situations, whether it is designing a website, setting up and customising a range of hardware and software, analysing systems or just having an awareness of the impact that communication systems have on industry.

You do not have to have an extensive knowledge of ICT to embark on the BTEC National qualification. Each of the units provides a good coverage of the subject matter. In conjunction, this accompanying book provides additional support in terms of a range of activities, case studies and test-your-knowledge sections alongside more comprehensive information that follows the guidelines of the specification.

The range of units available on the BTEC Nationals IT Practitioner qualification is quite diverse. The units provide opportunities for you to study at a very specialist level focussing on communication and interpersonal skills, hardware support and communications, web design, systems analysis, systems security and software design.

On successful completion of this qualification the progression opportunities are quite varied, you could progress onto a Higher National Diploma, Foundation Degree or a Degree programme. Alternatively, you could apply for careers in the areas of hardware support, web design, programming, end user support or other roles that have a business or ICT element.

How to use this book

This book provides a support mechanism for the BTEC Nationals IT Practitioner specifications. A range of core and specialist units have been covered within the text, each chapter providing a range of additional materials, activities, case studies and test your knowledge sections.

Each chapter begins with an overview of the content of the related unit and addresses the learning outcomes. Following on from this each of the main headings provides detailed coverage of the learning outcomes.

The activities have been designed to establish your level of learning and provide further opportunities for you to develop your understanding of a specific topic area or concept. The activities are devised to be used at an 'individual', 'group' or 'practical' level. The activities are broken down into a range of tasks that require you to undertake research, develop an understanding, provide an opinion, carry out an activity, discuss and present information.

The question sections provide you with an opportunity to re-visit and refresh your understanding of a previous topic.

In some areas of the book certain terminology is used that you may be unfamiliar with. To support your understanding of this, sections have been included that provide clarification or a definition of the terms referred to.

Courtesy of iStockphoto, diane39, Image# 4313218

Communication at all levels is essential in today's working environment. The ability to exchange ideas, discuss progress, update and provide feedback, set goals and objectives, plan and make decisions both on an individual, team and organisational level is paramount. If a company is going to grow and be successful within the marketplace they have to know and maintain their target audience and be more strategically focussed and competitive within their business environment.

Chapter 1

Communication and Employability Skills

This chapter will provide you with an insight into what employers are looking for in terms of a well-rounded employee in today's society. The skill requirements of the past have now been overtaken by a range of softer, more transferable skills and knowledge.

The need for good, transferable skills and the ability to be an effective communicator are valued very highly among employers, and it is these types of skills that individuals should be developing. Within the chapter you will be introduced to a range of attributes that employers value. In conjunction with this, you will learn how to communicate effectively and the contribution that ICT makes. Finally, the chapter will explore the area of personal development plans and ways of addressing them.

Each section within the chapter will focus on one of four learning outcomes:

- Understand the attributes of employees that are valued by employers.
- Understand the principles of effective communication.
- Be able to exploit ICT to communicate effectively.
- Be able to identify personal development needs and the ways of addressing them.

Embedded within each section there will be a range of activities that will strengthen your understanding of the subject matter and provide you with the support you need to complete effectively some of the evidence requirements for the unit.

Understand the attributes of employees that are valued by employers

Employers today want it all: knowledge, skills, technical and practical ability and above all, it could be argued, a good portfolio of transferable skills. The ability to communicate effectively within the workplace seems to be increasingly a highly desirable attribute.

Employers demand a great deal from any potential employee. Usually, job applications and person specifications identify essential and desirable skills, with many of the essential criteria including:

Specific job-related skills	✓
Technical knowledge	✓
Planning and organizational skills	✓
Effective time management skills	✓
Ability to work within a team	✓
Good written and verbal skills	✓
Good numeracy skills	✓
Having the right attitude	✓
Making a good contribution to the organization's aims and objectives	✓

To be successful in your chosen profession and job role it is important to develop a portfolio of these skills which can demonstrate that you are the right person for the job in today's competitive market.

Specific job-related attributes

Job-related attributes are very specific to a particular organization or role. The need for these attributes may change depending on the functional department you work for, your position and your seniority.

The need for technical knowledge is very person and job specific. For example, a network administrator may need to know about:

- network topologies
- systems architecture
- network protocols
- how to configure systems hardware and software
- network security procedures
- performance monitoring,

whereas a project manager would need a different skill set based on:

- resource planning and allocation
- target setting and monitoring
- time management
- budgets and financial control
- specific project management software knowledge.

Technical knowledge can be learnt and enhanced through training and development to ensure that an individual is equipped with the skills and ability to perform their job role.

Activity 1.1

For each of the following job roles identify two desirable skills and one essential skill that they may need.

Job role	Desirable skills	Essential skill
Lawyer	1	1
	2	
Teacher	1	1
	2	
Electrician	1	1
	2	
Store manager	1	1
	2	
Doctor	1	1
	2	

Technical knowledge

Technical knowledge will provide you with the specific theory and skills required to perform your job role. Technical knowledge and expertise will vary depending on the job role; for example, technical knowledge required to be a surgeon may include the study of anatomy, surgical procedures, and knowledge about medicines and anaesthetics. Technical knowledge for an accountant, however, may include financial accounting procedures, auditing techniques, preparation of company accounts, and profit and loss techniques.

Working procedures and systems

In conjunction with technical knowledge individuals also need to be aware of working procedures and systems. Working procedures and systems can be classified as generic or specific. Generic implies that all organizations have similar systems in place to address issues such as health and safety, data protection, and compliance with other legislative and statutory requirements. Therefore, an awareness of the Health and Safety at Work etc. Act 1974 will be a standard procedure within most organizations. Specific working procedures and systems are unique to the organization and these could include compliance with a set of rules, codes or professional practice. For example, a specific policy may have been set up regarding dress code.

Working procedures and systems in most cases are not taught or studied, unlike the requirement for technical knowledge. These procedures and systems require an awareness and sometimes compliance by the individual once they are established in their job role. Working procedures and systems are essential to any job role as they provide the guidelines and instruction in terms of working and behaving as a professional.

Activity 1.2

Choose an organization, or an individual who works in an organization, that is easily accessible to you.

1. Produce a questionnaire or an interview sheet with at least five questions. These questions should include identification of a range of technical knowledge and working procedures and systems.
2. Present the interview sheet or questionnaire to the organization or individual and ask them whether they can respond to the questions NB. The questions should identify what technical knowledge is required to enable the individual or employees within the organization to meet the requirements of their job role. In addition, responses should be given with regard to the types of working procedures and systems that are in operation within the organization.
3. Once this activity is complete, the group should come together to share the information obtained. The group should discuss the variations and similarities in technical knowledge required and also examine any variations in working procedures and systems.

General attributes

Contributions made by the individual help to develop the role that they are in; therefore, certain general attributes are required to ensure that a job role can be carried out effectively. These attributes may include:

- planning and organizational skills
- time management
- team working
- verbal and written skills
- numeric and creative skills.

When you apply for a job, the requirements listed in the job description or job role usually refer to a range of other skills in conjunction with technical knowledge. These other attributes may demonstrate how an individual can plan and manage their time, interact within a group and communicate effectively, demonstrating that they are numerate and also creative in their thinking and application to tasks.

Planning and organizational skills are important because they demonstrate that an individual can organize, anticipate and manage a range of activities, which is very important in a job role.

Example   A job role may involve you working as part of a project team where you have a responsibility to research and select a venue choice for a conference.

Your planning and organizational skills may include:

- researching and visiting different venues
- selecting a venue

- planning the resources required to support the conference – transportation, food, delegate agenda for the day, accommodation, guest speakers, etc.
- organizing different resources to ensure that everything goes to plan on the day.

Linked in to planning and organizational skills is also 'time management' – ensuring that any tasks that you undertake fall within the boundaries of a specified time.

Example   As part of your job role you are expected to select the venue and organize the conference event schedule 21 days before the conference and send information packs out to all of the delegates.

Working within a team is very important as the ability to communicate and function with other colleagues, third parties and stakeholders may be critical to your job role. Working within a team can be essential for some elements of your job role in terms of pooling resources, knowledge and skills and working towards a common objective or task.

Example   Within your project team you are expected to have weekly meetings to discuss the conference developments and update fellow team members about the progress that you have made and any issues that you have identified that could impact on their area of the project.

The ability to communicate skills and knowledge verbally in a written format, numerically or creatively is also essential to any job role.

Example   Part of your job role working within the project team is to report back on the progress made in terms of searching for a suitable conference venue. At the weekly team meeting you have been asked to produce a brief written update that can be distributed to the team outlining current developments. In addition, you are expected to verbally discuss venue options. In terms of applying numerical and creative skills, identifying the costs involved in running the conference event and creating the delegate brochures to support the conference are further examples of the need for a range of general attributes that are required to support individuals in their job role.

Attitudes

Adopting the right attitude is essential in all areas of professional practice. All individuals have their own different attitude styles; some people are quite relaxed and informal in their outlook, others are more formal; some people generally have a positive approach and attitude, some more negative.

In a professional context your attitude could change depending on your job role, individual tasks for the day, the people with whom you are working and the environment in which you are working.

Activity 1.3

1. For each of the attitudes listed in the table, provide an example of how this could be shown or reflected.

Attitude	Example
Determination	
Independence	
Working with integrity	
Tolerance	
Being dependable	
Problem solving	
Leadership	
Confidence	
Self-motivation	

2. How would you describe your general attitude to things and what things have an impact on this?

Organizational aims and objectives

An organization can have a range of aims and objectives that are linked to various targets, markets, products/services and resources. The aims and objectives of an organization can also be linked to meeting the expectations of employees, customers, shareholders and a range of other stakeholders.

The aims and objectives of an organization are also linked to their 'mission statement' as they provide the framework for achieving short and long-term goals.

Examples of organizational aims and objectives are:

- to be more competitive within the marketplace
- to achieve a larger market share
- to create a stronger brand name
- to diversify into global markets
- to engage more with e-commerce
- to be more profitable.

Mission statement – defines the main purpose or function of an organization. It provides a declaration of an organization's goals, values, beliefs and aspirations.

Although there are many aims and objectives within an organization, they are not realistic or feasible unless specific targets and measures are provided as a framework for achievement. In terms of setting a framework, aims and objectives should be time related in terms of in the short, medium or long term (Table 1.1).

Table 1.1 Framework for setting and meeting aims and objectives

Classification	Period	Example of aim/objective
Short	1–3 years	Establish a strong brand name within the UK market
Medium	3–5 years	Diversify into European and global markets
Long	5 years+	Become a world market leader

Understand the principles of effective communication

Effective communication implies that the information that has been transmitted from a sender to a receiver is clear, fit for purpose and timely. A sender or source of information could be an employee, a friend, customer, student or teacher, for example, and the receiver or recipient could be a single person, multiple people or a target audience of some description.

The way in which information can be communicated effectively can also depend on the way in which we communicate in terms of general communication skills and interpersonal skills. Effective communication is also very reliant on the transmission tool or method of communication; for example, spoken format, electronic, visual or written format.

General communication skills

To engage in effective communication, a number of points need to be considered that look at the way in which information is delivered and the techniques used to capture the interest of an audience. For some, trying to communicate effectively can be difficult, especially with a wide range of barriers such as geography, limited communication resources, time, and suitability and choice of methods. However, imagine how difficult it must be to communicate with somebody who originates from a different country with a different language, customs or culture.

Cultural considerations are very important when you are trying to engage in communication. A different culture could mean a different part of the country or world, a different language or accent, different beliefs and customs that could be politically, socially, environmentally or religiously based. Culture and communication is very important because it could impact on how information is delivered and when it is delivered to ensure that you are taking into consideration and respecting the culture of your target audience.

An example of this is 'greetings'. In one culture it may be perfectly acceptable to embrace and possibly kiss a person as a sign of welcome. In some cultures a kiss on both cheeks may be customary; however, in other cultures to embrace somebody physically, especially a woman, may be deemed immoral and disrespectful of that culture.

Adapting the style and content of communication and tailoring your delivery to a specific audience may also be a way of ensuring that you are engaging in effective communication. Modulating your voice so that it has a certain rhythm in the delivery, adjusting the tone and speed, emphasizing certain words or phrases, and using appropriate terminology can all increase the effectiveness, in conjunction with the delivery format.

Ensuring that the information is accurate and has been generated from a reliable source are also important factors in communicating effectively. If out-of-date or inaccurate details are provided this could have a negative impact on future decisions and planning proposals in the short and long term. What would happen if a market leader claimed that they had the lowest price for a certain product, when in fact they did not?

Differentiating between facts and opinion can be difficult, especially if you are trying to communicate an issue that is of a sensitive, moral, ethical, personal or emotive nature. However, it is important that facts are stated as being based on research and evidence, whereas opinions are usually based on experiences or hearsay.

Another challenge is to engage and keep an audience interested in a topic of conversation, especially if the audience is large. There are several ways in which you can try to engage with an audience, for example changing the speed and tone or your voice and using emphasis to stress certain key points. In addition, using a range of visuals or props, such as an interactive whiteboard or electronic presentation materials with built-in animation or multimedia tools, may support the message that you are trying to convey and capture the attention of the audience.

Question-and-answer sessions can also be used to identify how effective communication is or has been. Questions can stimulate the audience into thinking about what information has been communicated and/or messages conveyed, and then provide feedback in the form of an answer. Moreover, questions may be raised by the audience that require additional output from the presenter.

Interpersonal skills

Interpersonal skills and the development of interpersonal skills are crucial in all aspects of social, personal, educational and professional environments. Being able to communicate with a person or people at all levels is very important, and how you communicate information in terms of the delivery method, style and use of body language reflects on the type of person you are and impacts on the effectiveness of your interpersonal skills.

Methods for communicating interpersonally

A range of methods can facilitate interpersonal communication, including:

- verbal exchanges
- signing
- lip-reading.

Verbal exchanges

Verbal exchanges can cover a whole host of situations where verbal dialogue is required, as shown in Figure 1.1.

Figure 1.1 Example of verbal exchanges
(Source: Microsoft Clipart online)

Verbal dialogue is important because it is a direct form of communication and the response is received almost immediately, especially if the verbal exchange is conducted in person, not over the telephone, etc.

Signing

Signing or sign language is an effective way of communicating with a person who has a hearing impediment and cannot hear any vocal dialogue; it can also be used by people who suffer from aphasia. Sign language uses body language, hand movements, facial expressions and lip movement to convey words and thoughts. Figure 1.2 provides an example of how letters are conveyed using hand movements.

Aphasia – a condition that restricts the ability to speak and/or comprehend any language owing to brain damage. Aphasia can develop as a result of injury or can be from birth. There are various degrees of aphasia; some people may be able to speak but not to write, while others may not be able to pronounce or form words.

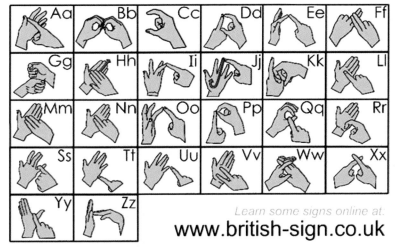

Figure 1.2 Signing alphabet
[Source: Waterfall Rainbows, http://www.british-sign.co.uk/fingerspelling_chart.php].

Lip-reading

Lip-reading provides an effective form of communication for people who have a hearing impediment or a condition such as aphasia. Words and phrases can be read from the movements in the mouth, face and lips.

Techniques and cues

A range of techniques and cues falls under the category of interpersonal skills. These include:

- use of body language
- use of intonation
- use of 'smileys'
- capitalization in text in e-mails.

Body language can be a very powerful and effective way of communicating with people face to face. The use of hand, arm, face and body movements to make a stance or gesture can have quite an impact.

Activity 1.4

1. Write down all of the different ways you have used body language over a 24 hour period and identify when, where and to whom you expressed body language.
2. Do you feel that your body language changes depending on the mood you are in? Why do you think this is?
3. For each of the emotions and situations shown in the table, provide an example of suitable body language expression that could be used.

Emotion/situation	Body language example (s)
1 Greeting a friend	
2 Trying to grab someone's attention who is 20 metres away but facing you	
3 Excited because you have just won a prize	
4 Feeling scared	
5 Feeling happy	
6 Feeling sad	
7 Feeling angry	
8 Feeling nervous	

Intonation means that there has been a change in the tone of someone's voice; this could be a variation in pitch with the possible introduction of irony. Intonation can be classified as 'rising' where the pitch of the voice increases over time or 'falling' where it decreases over time. Intonation can also be referred to as 'peaking' where the pitch rises then falls, or 'dipping' where it falls then rises.

Smileys ☺ ☹ are symbols that express how you are feeling without actually stating it. They are used widely on the Internet, and in messenger services, chat rooms, blogs, forums and texts.

Capitalization of text in e-mails can also be used to express and emphasize certain key points or phrases of interest, thus representing another form of interpersonal communication.

Positive and negative language

Language can be deemed to be positive or negative depending on the way a phrase is constructed and delivered. The same sentence could be used both positively and negatively depending on the emphasis of certain words.

Example

- 'oh what a *lovely* hat!'
- '*oh* what a lovely hat!'

With the emphasis on 'lovely' this could be deemed as positive language. With the emphasis on 'oh' a hint of sarcasm has crept into the same sentence and this can now be deemed as negative.

Paying attention and active engagement

It is very important when using interpersonal communication tools and techniques to ensure that you have the audience's attention and that they are actively engaged. There are several ways of checking that this is taking place. Maintaining eye contact can help to keep the attention of a person or audience; if they feel that you are addressing them as an individual they may be more responsive. Telltale signs such as nodding and smiling in agreement or frowning in disagreement are also ways of observing whether or not the person is engaged. General fidgeting, tapping of fingers or toes, or vague expressions from a person or audience may indicate a lack of engagement and disinterest.

Understanding barriers

Various barriers can impact on interpersonal communication, including background noises, distractions and lack of concentration. If there is a background noise present when you are trying to communicate with somebody this can pose a barrier to communication; for example, an open window with traffic or roadwork noise could block out any message that you are trying to convey. Other distractions such as interruptions, fidgeting and the presence of other people can also act as barriers. A lack of concentration could mean that important elements of a conversation are lost or misinterpreted.

Questioning

Many types of questions can be used to establish facts, gather feedback, assess topics that have been learnt and clarify certain points. Questions can be categorized as being 'open', 'closed' or 'probing'.

- **Open question** – inviting an open and detailed response:
  - What do you feel are the most important qualities valued by an employer?

- **Closed question** – inviting a yes or no response:
  - Are interpersonal skills important?
- **Probing question** – inviting a detailed response, possibly based on personal opinion or judgements:
  - Interpersonal skills are valued highly by employers. What do you think about this?

Activity 1.5

You can question people in a number of ways, and the type of question will have an impact on the nature of the answer – whether it will be descriptive and lengthy or short and precise.

1. Devise a questionnaire that is aimed at one of the following topics:
   - sports
   - film and television
   - musical tastes.
2. Within the questionnaire include at least three (of each) open, closed and probing questions.
3. Give the questionnaire to five people to complete.
4. Analyse the results of the questionnaire to see what types of answers have been given and whether there is any relationship between the answers and the questions asked.

Appropriate speeds of response

The response time will depend very much on the type on question, issue, information that has been requested and the form of communication used.

If you extend your arm to shake hands with another person the response should be almost immediate by the recipient. If you are engaging in conversation then the response should be quite dynamic, especially if you are talking face to face.

Communicate in writing

Written communication is a way of confirming issues that have been raised and assurances and guarantees that have been made. Written communication can be quite formal in terms of providing a legal seal such as a contract of employment or a business agreement. Written communication can also be quite informal in terms of writing an e-mail or a letter to a friend.

Following organizational guidelines and procedures

All organizations have a set of organizational guidelines and procedures that are used to ensure compliance with various statutory and regulatory systems and to promote a safe and professional working environment. The need for written organizational guidelines and procedures is paramount to ensure that there is a permanent reference source that can be accessed by both employees and employers within the organization. In addition, written guidelines and procedures can easily be viewed and communicated to ensure that everybody within an organization understands their own responsibilities and those of others.

Identifying and conveying key messages in writing

Written communication is very important in terms of conveying key messages. There are various ways that information can be communicated in writing, for example letters, faxes and e-mails.

Letters can be classified as being formal or informal, depending on the content, sender and proposed recipient. A letter to a friend enquiring about their well-being is classed as being informal, whereas a letter to a potential employer accepting the offer of a job would be considered to be formal.

Most letters, especially formal ones, tend to follow a set template for design that includes some form of 'identifier', an 'introduction', 'content' and some form of 'closure', as shown in Figure 1.3.

IDENTIFIER
Sender's Information:

> Mrs Patricia Samms
> 13 The Guild
> Clayborn Road
> Hockering
> NR17 3PD
>
> _____
>
> If there is no letterhead the address could go to the right hand margin:
>
> Mrs Patricia Samms
> 13 The Guild
> Clayborn Road
> Hockering
> NR17 3PD

Recipient Information

> TEY Archaeological Group
> 4 Hinders Chase
> Pastor Road
> Norfolk
> NR13 5SY
>
> Date
> 24 March 2007
>
> Reference Number: (if applicable)
>
> Salutation: For the attention of Dr Spencer James
>
> **INTRODUCTION:** Discovery of Roman Coin circa. 66 AD
>
> **CONTENT:**
>
> Dear Dr James
>
> Following our recent telephone conversation regarding my find, which I suspect is a Roman coin circa 66 AD, please find enclosed a description of the coin including some visible marking and full details regarding my discovery.
>
> **CLOSURE:**
>
> Yours faithfully
>
>
> Patricia Samms
> Enc.

Figure 1.3 Sample letter

Identifier – provides identification as to who the letter is for and from whom it has been sent.

Introduction – provides a brief summary, possibly a line or two to identify the proposed content of the letter.

Content – this is the body of the letter where all of the information is placed regarding the topic or theme.

Closure – this completes the letter by providing a name, a signature and any reference to other documents to be sent with the letter, such as a curriculum vitae or receipt. You can identify whether other documents have been enclosed by the following 'Enc.' at the end of the letter.

Faxes or facsimiles are documents that are sent using some form of ICT media, for example a fax machine or a computer that has fax capabilities. A fax can also have quite a defined format, as shown in Figure 1.4.

Fax

To:	From:
Fax number:	Pages: of
Contact number:	Date:
Re:	CC:

Urgent ☐   Review ☐   Response required ☐   No action required ☐

CONTENT:

Figure 1.4 Sample fax

E-mails seem to be a preferred method of communicating in today's society. Information that is usually sent in a letter can now be reproduced in an e-mail and sent a lot more quickly and more cost-effectively. In addition, one e-mail can be sent to many people and a response can be almost immediate (Figure 1.5).

Figure 1.5 Sample e-mail

Using correct grammar and spelling

Written communication provides a clear and structured method of transferring information between people. However, the clarity and structure of written communication can be impeded if the content of the information sent is not grammatically correct in terms of spelling, punctuation or sentence construction. Depending on the format of the written communication various checks can be enforced to ensure that errors of this nature are limited. If the written document has been produced electronically using word-processed software, spelling and grammar tools can be used to track and amend any errors. If the document has been produced by hand, then proofreading checks need to be carried out by the sender.

Structuring writing into a logical framework

Written documentation can vary in terms of formality, content and layout. Some written documents have a very prescribed template, as shown with the letter, fax and e-mail templates. Other documents can be created with a more freestyle approach.

Identifying relevant information in written communications

All communication can be classed as relevant in some way, otherwise what would be the purpose of communicating it? Some information, however, especially in a written format, can be more relevant than others. For example, in a written contract there will be a great deal of relevant information regarding terms and conditions and legal responsibilities and obligations. In a letter to a friend giving an invitation to a party, the relevant information may include the date, time and venue.

Relevant written information can be identified in a number of ways, including:

- a change in font style or size
- use of bold, underlining, capitals or italics
- indenting a word, sentence or paragraph
- highlighting text in a different colour
- applying some form of formatting – using a border, numbering or bullet function
- using a comments box
- applying a heading or subheading.

Reviewing and proofreading your own written work

The need to review and proofread written work is essential to ensure that the message conveyed is clear and logical, and meets the requirements of the target audience.

Depending on how a written document is created, this process can be automatic in terms of software tools identifying errors and providing a facility to correct them. In conjunction, reading through any written work to ensure that it makes sense and reads appropriately is also essential prior to any written document being sent or submitted.

Conveying alternative viewpoints

Sometimes it is important to convey a number of different viewpoints within a document. Examples of this are:

- completing an assignment that examines different hardware solutions for an organization
- identifying the best piece of antivirus software available on the market
- researching current trends and justifying the buying patterns of 16–19-year-olds.

By introducing a range of alternative viewpoints your audience can then make a decision or a conclusion can be reached that is based on a range of evidence and not just a single response or a personal opinion that could be biased.

Reviewing and editing documents created by others

Several methods can be used to review and edit documents that have been created by another person or other people. These review and editing techniques can range from general proofreading to checking for currency, correctness and relevance in the content. Other techniques may include highlighting text, using comment boxes, inserting footnotes or endnotes or applying some sort of formatting such as 'strike through' to delete text.

Note taking

Note taking is quite an important feature of written communication as it can form the basis of future documents. There are many ways in which notes can be taken (Figure 1.6).

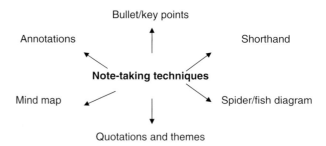

Figure 1.6 Note-taking techniques

- Bullets and key points: taking notes using these methods is very quick, specific and easy to apply, especially with the built-in bullet and numbering features of application software packages. Information in this format can be condensed in single keywords or short sentences. An example of this can be seen in the following extract:
  'Issues to consider when setting up a website:
  − costs
  − hardware
  − software

- connection
- ISP charges
- maintenance and service charges
- security
- design.'
- Shorthand: unique styles of scribing complete words into short symbols, to enable faster note taking.
- Spider/fish diagram: a method of taking notes visually using keywords and word or subject associations.
- Quotations: writing down specific phrases or quotations from particular authors or sources that can be used as references for future research.
- Mind maps: Buzan (1989) developed a style of transferring notes into a diagrammatic format known as a 'mind map' or 'brain pattern'. The core of a mind map starts with a word or phrase in the centre of a page which represents the main theme; branches then extend outwards incorporating links to that theme, and again into subthemes.
- Annotations: these are used to provide an expansion to thoughts or ideas. Annotations can be textual or graphical.

Be able to exploit ICT to communicate effectively

ICT can be used to facilitate effective communication. It can ensure that information is communicated professionally, effectively and efficiently. By exploiting ICT, communication can also be targeted towards a specific audience or multiple audiences and presented in a consistent and professional layout. ICT can also be used to monitor that information has been sent and received and that any appropriate action or response is provided within a short timescale, in contrast to waiting for postal confirmation.

Communication channels

A number of communication channels can be supported by ICT tools and software. These include:

- word-processed documents
- presentations
- web pages
- e-mail
- specialist channels such as blogs, vlogs, podcasts and video conferencing.

Word-processed documents

The use of applications software to create various documents can be of great value, benefiting both home and business users. One of the most common and varied sources of documents created is that of word-processed documents. Using word-processing software, a range of documents can be created, including:

- letters
- invoices

- statements
- invitations
- menus
- leaflets and brochures.

Word-processed documents can feature a range of desktop-publishing features, editing tools, graphics and formatting capabilities to customize and enhance the generated text.

Activity 1.6

Create two different types of word-processed documents. The documents should be tailored to meet the needs of two different end-users and incorporate a range of editing and formatting tools. One of the documents should also contain a graphic.

Presentations

Presentations are an excellent way to deliver visual material to a target audience. Presentations can easily be designed with the help of ICT tools using individual slide designs or templates to deliver text, graphics, multimedia, web links and other required information.

Web pages

People set up and view web pages for a number of reasons, all of which embrace effective communication.

Web pages can be used to provide general information, such as travel timetables or cinema viewings, and create an awareness of certain topics or issues. Web pages can be used to book, buy and sell products with the use of e-commerce facilities such as Ebay or Amazon. Web pages can be used to engage people in topics of conversation and debates through the use of forums, discussion groups and mailing lists. Finally, they can be set up for entertainment purposes to provide music, video and downloaded clips of multimedia snippets.

E-mail

E-mail can be considered to be a very effective form of communication, because the principle of typing a message and pressing a button to send it to another person is quite straightforward.

There are many benefits associated with e-mail and its effectiveness as a communication tool, as identified in Figure 1.7.

E-mail encourages effective communication because of the way it is designed to make things simple and cost-effective. One e-mail can be sent to one person or multiple people, copies can be sent, there are no associated postage costs, a response can be almost immediate and the design of the e-mail can be customized.

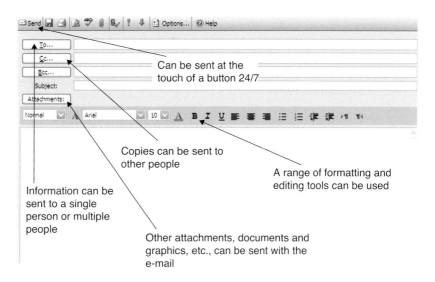

Can be sent at the touch of a button 24/7

Copies can be sent to other people

A range of formatting and editing tools can be used

Information can be sent to a single person or multiple people

Other attachments, documents and graphics, etc., can be sent with the e-mail

Figure 1.7 E-mail benefits

Specialist channels

A range of specialist ICT channels, such as blogs, vlogs, podcasts and video conferencing, can facilitate effective communication.

Blogs or web logs are journal-style diary entries that are set up on a website for others to read. Blogs are created chronologically and provide accounts of personal, social, professional, political or other topic areas. Blogs are quite dynamic in that people reading through the entries can sometimes leave their own views and opinions. Vlogs or video logs have the additional feature of video footage that can be used to give a multimedia aspect to any entries.

Podcasts have become extremely popular recently, with a whole host of personalities broadcasting messages to their fans and captive audiences, including Radio One DJs, Ricky Gervais, and The Queen with her Christmas Day message.

The function of video conferencing is to transfer and broadcast messages, presentations and seminars, and to hold meetings between people who are based in different locations.

Activity 1.7

A range of specialist channels exists today, including blogs, vlogs and podcasts.

1. Select one of these three specialist channels and carry out research to address a range of elements (bulleted).
2. Select three different sites/broadcasts within your chosen specialist channel and produce an information sheet or a booklet that looks at:
    ● the topic and author
    ● the content

- how it is displayed – layout, colour, graphics, multimedia components, etc.
- information value
- appeal to your age range.
3. Present your findings to the rest of your class.

Benefits and disadvantages

There are many benefits associated with the use of various communication channels to communicate effectively.

Current technology provides numerous opportunities to enhance communications in terms of the way in which it is presented and delivered, the speed of delivery, clarity and audience extent. For example, an e-mail can be composed in a specific format, another document can be attached, and it can be sent instantly to a single person or a range of people.

Although the benefits of using specific communication channels are wide-reaching, there are also some disadvantages. The disadvantages of using ICT as a communication channel include:

- **Initial costs** – hardware and software may have to be purchased to initiate communications, especially in the case of using e-mail.
- **Difficulty of use** – some people find it difficult to interact with technology, therefore it may be far easier for somebody to write and post a letter than to learn how to use e-mail and send accordingly.
- **Unreliability of communication channels** – sometimes there may be situations where the associated hardware or software is not functioning properly, which means that a presentation cannot be delivered or an e-mail sent.
- **Incompatibility** – if a document, presentation or e-mail has been created using a certain piece of hardware or software, different hardware or software used to complete or send the document may be incompatible, rendering communication ineffective.
- **Security issues** – sometimes there are issues concerning the security of certain communication channels, especially if preventive steps have not been taken to encrypt the information

Software

Software can provide a range of opportunities in terms of facilitating effective communication. Software can be categorized in terms of its functionality, for example:

- operating system software
- applications software
- specialist/customized software
- utility software.

Operating system software interacts with the hardware of a computer to ensure that the system resources are managed, controlled and coordinated. Operating system software is used on both standalone and networked systems and can be described as being text or graphical based.

An example of a text-based operating system is that of MS DOS, and graphical-based systems include Windows XP and Vista.

Applications software allows users to carry out information processing activities such as:

- word processing
- numerical and financial modelling and statistical analysis
- design and desktop publishing
- document management, presentation, storage, retrieval and manipulation.

Applications software also enables users to interact with hardware through different multimedia and communicative techniques.

Specialist and customized software is designed to support a specific function or target audience. Specialist software includes software designed to meet the requirements of an end-user, for example customized management information systems (MIS) software, or software to support a user who may have a specific job role or disability.

Utility software provides the tools to support the operation and management of the system, for example:

- virus checkers
- security software, e.g. firewalls
- defragmentation software
- partitioning software
- CD authoring
- DVD playback software.

Word-processing software

Word-processing software is probably one of the most common pieces of software used because of the range of flexibility and functions that it offers.

Word processing allows you to compose a written document such as a letter, report, story or paper very easily. Once a new page has been selected and a font style and size have been chosen the document begins, just like using a typewriter but with the added benefits of editing and saving material as you go along, without the need for correction fluid. Word-processing software also allows you to engage with a range of tools and techniques that enable you to add or import graphics, format and customize pages, provide specific design templates and interact with other software applications. Figure 1.8 provides an example of how word-processing software can be used.

Presentation software

Presentation software can be used to create and deliver professional presentations. Presentations can be built up on individual slides (Figure 1.9) to create a slide show that can be presented slide by slide at the touch of a button or automatically timed to be delivered as a single slide show.

Dessert

Lemon tart with clotted cream
Berry crumble with vanilla custard
Fresh fruit salad
Warm chocolate soufflé

Figure 1.8 Use of word-processing software

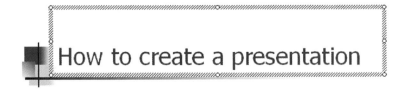

Using presentation software

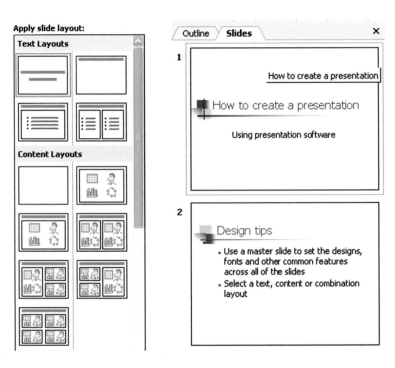

Figure 1.9 Presentation design, tools and techniques

Many tools and techniques are available to change the font style and size, and the slide design and layout. A range of transition techniques can also be used to introduce each slide, for example text dissolving, fading in and out, or flying in from a particular side.

Other software

A range of other software can be used to communicate with other people; one example of this is e-mail. E-mail software allows you to create a message or send an attachment to a single user or multiple users. E-mail software includes standard tools that allow you to send and receive messages, sort messages in an inbox, save, edit and delete messages. Some e-mail software is solely for the purpose of communicating messages, such as 'Eudora' and 'Outlook Express', while other e-mail software is embedded within more advanced personal information management systems that include appointment scheduling, calendars and notes.

Some software is specifically designed to meet the needs of people with special needs, requirements or disabilities. A good example of this is software designed for people who are visually impaired. The types of software available to support visually impaired users include:

- Screen readers that read aloud information displayed on a monitor or screen. This can include text within a document, information within dialogue boxes, error messages, menu icons and graphical icons on a desktop. Examples include JAWS, outspoken and KeyRead.
- Screen magnification software that enlarges the information on the screen, such as Supernova and ZoomText.

Software tools

Software tools are available that can support users in creating documents and checking documents for consistency and completeness, and ensuring that they are presented in a professional format. Some of these tools can be referred to as 'proofing' tools, such as thesaurus and spellchecker. These tools allow you to check the spelling, grammatical construction and meanings of words.

The thesaurus tool allows you to check the meaning of a word and also find a replacement word that means the same, for example the word 'start' as shown in Figure 1.10.

Spellcheck allows you to check the spelling of a word. This is usually identified by a wavy line underneath the word. To correct the spelling you can right-click on the word, or go through the 'spelling and grammar' options on the 'tools' menu as shown in Figure 1.11. If the function has been enabled automatic correction can also be initiated.

Other tools are also available within software packages, for example conversion of tabular information into graphics and also text readers.

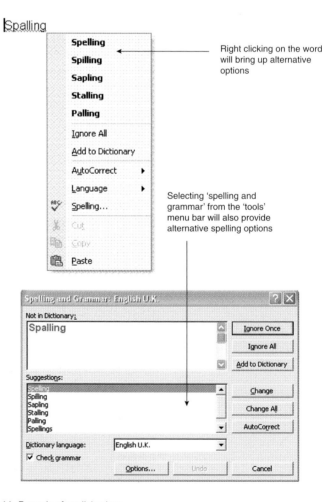

Figure 1.10 Example of thesaurus

Right clicking on the word will bring up alternative options

Selecting 'spelling and grammar' from the 'tools' menu bar will also provide alternative spelling options

Figure 1.11 Example of spellchecker

Be able to identify personal development needs and the ways of addressing them

Personal development needs can be identified in a number of ways and through the use of different resources. Within an organizational context it is important to highlight the needs of employees and ways in which these needs can be addressed. By addressing these needs an individual may become more motivated and productive in their job role, and they could develop new skills or strengthen existing skills, or qualify for a more challenging role.

Identification of need

Identifying a personal development need can be challenging. You need to be aware that the need exists, and sometimes this is only recognized at a set time or periodically following an appraisal or a self-assessment, or through somebody else such as a colleague or peer. Needs can also be identified by customers in terms of providing feedback on a particular product or service, or through performance and target data.

Self-assessment

Self-assessment is a term used for the review of individuals, assessing a particular need, criterion, task or requirement. Through self-assessment individuals can reflect on what they have achieved or completed over a certain period, record this formally and then use this to address certain needs or requirements, such as training or development.

Formal reports

Formal reports are another way of identifying and capturing the needs of individuals, and are especially effective following an appraisal. The formal report will outline various targets that need to be achieved, in conjunction with objectives and dates for achievements. The report may also identify in detail how these needs are going to be addressed through training and staff development. It could propose specific courses and training events required to be undertaken throughout the year, for example.

Other methods

Needs can be identified in many other ways. Two further approaches include the use of customer feedback and the use of performance data. Customer feedback is a good way to capture information in terms of opinions and satisfaction levels, and can be obtained through:

- product or service questionnaires
- surveys
- interviews
- tick sheets or 'happy sheets' (Table 1.2)
- focus groups
- question-and-answer sessions.

Table 1.2 Example of a happy sheet

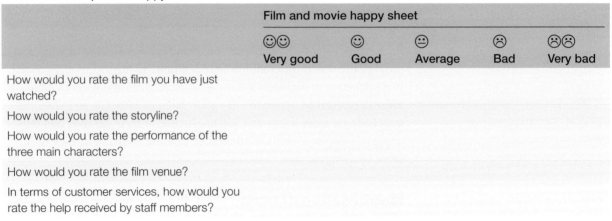

	Film and movie happy sheet				
	☺☺ Very good	☺ Good	☻ Average	☹ Bad	☹☹ Very bad
How would you rate the film you have just watched?					
How would you rate the storyline?					
How would you rate the performance of the three main characters?					
How would you rate the film venue?					
In terms of customer services, how would you rate the help received by staff members?					

Customer feedback can be obtained manually through postal surveys or distributed questionnaires. Feedback can also be obtained dynamically by interviewing customers, or online through an organization's website. Online feedback mechanisms are very popular, and usually easy to complete and send, at the touch of a button.

Performance data can be useful to establish areas and needs such as:

- productivity and efficiency levels
- targets achieved and not achieved
- products, stock and services sold or not sold.

Performance data is useful because it provides guidance and measures in terms of what, when, how and where. Performance data can be used to track and monitor stock, provisions, services, resources and people, and in conjunction with an MIS, the storage and manipulation of this data is far more effective in providing reports and other useful information to make an organization more cost-effective, productive, profitable or competitive within the market.

Records

To make personal development planning effective there should be some sort of mechanism or audit trail that can be used as a reference throughout the period of development. In addition, an individual may want to reflect on what they have done in the past, to improve in the future. There are several ways in which development planning can be recorded, such as specific personal development plans (PDPs), determining where targets can be set, and through the use of appraisal documents.

Personal development plans are tailored to address and meet an individual's needs. They can set out targets and identify ways in which these targets can be met, and over what timescale.

Target setting is very important as it can act as a motivator and also as a measure in terms of what you would like to achieve and by when.

Targets can be classed as being short, medium or long term depending on the nature and complexity of the target and other associated resource factors. Targets should also adhere to a specific framework if they are to be effective; this framework ensures that targets are SMART:

- specific
- measurable
- achievable
- realistic
- time-framed.

If targets do not address all of the areas of this framework it could be argued that the target set may become unachievable; for example, 'I would like to be the next Prime Minister' as a target may not be feasible because there is no indication as to how this will be achieved, how the target can be measured and within what time-frame: one, two, ten or twenty years? The target may be very focused and specific; however, it is unsupported in many other areas and it is also questionable whether the target is realistic, especially if the individual has no real interest in current affairs or involvement with politics.

Appraisal documents are usually prominent in organizational environments and are used to assess the performance and identify the needs of individuals. Appraisal documents can be used to motivate staff members by recognizing their achievements throughout a set period. These achievements could be related to a specific job role, such as helping to increase productivity by 15 per cent or excelling at a certain project. Appraisal documents can also be used to identify any areas that require further development and these are usually supported by action plans on how these areas of weakness can be converted into strengths. Finally, appraisal documents provide individuals with a mechanism to formally identify any issues, needs or requirements in their job role.

Methods of addressing needs

Once personal development needs have been identified they should be addressed to satisfy the requirement. Needs can be addressed in various ways, as identified in Figure 1.12.

Figure 1.12 Ways of addressing needs

Job shadowing effectively means observing another person in their job role and in some cases acting in their job role (possibly in their absence) to gain knowledge and understanding about what the role involves.

Attending courses and internal and external training are other ways in which individual and corporate needs can be addressed. From an individual's perspective attending a course may provide them with an opportunity to network with other professionals, to share good practice, and to pick up hints and tips on how to manage their job role better or how to achieve their targets. Attending a training course may introduce the individual to new experiences or provide them with opportunities to update existing knowledge and skills in a particular area. From a corporate viewpoint, sending employees on courses and training may indeed motivate the individual, making them more productive in their job role. In addition, information can be shared with other colleagues to update their skills and understanding as well.

Team meetings provide an opportunity for topics at all levels to be shared and discussed. Team meetings may be focused on a specific corporate or departmental need that can be shared with everybody in the meeting. Suggestions and recommendations may also be put forward from the meeting on how these needs can be addressed.

Formal events can provide an opportunity to meet a range of users' needs. Such events provide individuals with the opportunity to share good practice with colleagues and external third parties.

Activity 1.8

Recognizing a need can be difficult, especially if you are not aware that one exists. Needs can be identified by you, a friend, colleague, boss, teacher, parent or a range of other people.

1. Identify two needs that you have had in the past and describe:
   - what the need was
   - who identified the need
   - whether and how the need was addressed.
2. Identify a current need that you have and think about how you are going to address this need. Think about who or what else may be able to help you with this need.
3. What do you feel will be accomplished by addressing this need?

Learning styles

Learning styles can provide the answer to why different people respond in a certain way when working in a specific environment and with certain people.

Many studies have been carried out over the years on how people learn and why they learn in a particular way. These studies identify different traits and characteristics within individuals that form the foundation of a particular learning style.

Examples of systems

One learning style system categorizes learners into the following groups:

- active/reflective
- sensing/intuitive
- visual/verbal
- sequential/global.

Each group has its own set of characteristics and styles that defines the way in which individuals learn best (Table 1.3).

Table 1.3 Comparison of learning styles

Style	
Active	**Reflective**
Learn by doing – active learners, of a practical nature, the need to get involved. 'Let's get on and do it'	Learn through reflection – sitting back and thinking before doing. 'Think things through first'
Sensing	**Intuitive**
Using a method or a framework to address tasks and solve problems – very prescriptive, sometimes a repetitive learner	Use of more innovative methods. Dislike repetition and more into spontaneity. Like to discover new ways of doing things
Visual	**Verbal**
Visual learners work more effectively in a visual environment – through maps, drawings or pictures	Verbal learners are more effective with written tasks, instructions and guidelines
Sequential	**Global**
These learners progress by taking small steps	Global learners like to tackle tasks and problems as a whole

Identification of preferred learning styles

You can identify your learning style in a number of ways. Different studies provide guidelines and lists of characteristics to enable you to relate to certain attributes or a specific style. Tests can be taken that require you to answer a series of questions, which will determine the type of learning style that you have. These questions may require you to provide answers about how you work within a team and/or individually, your reaction to a particular situation, how you feel about a certain task or problem and how you would categorize yourself, for example as creative, a problem solver, a decision maker, an analytical person or a good planner.

Activity 1.9

Various paper-based and electronic learning styles tests exist, many of which can be completed online.

1. Find a learning styles test (e.g. Honey and Mumford), complete it and identify your own personal learning style.
2. Do you agree with the learning style category with which you have been identified?

How to benefit from knowing your learning style

It is possible that you have gone through your whole life not being aware of learning styles and what they can do for you. For some this is no real major disadvantage, but for others it can make a real difference to both their social and professional lives.

Knowing your learning style can bring many benefits; for example. learning styles can identify you as a person by the responses you have given. If you have been identified as having a particular style you can automatically see the strengths in your character. It could be that you are a very strong and independent person who works well as an individual and is self-directed in their learning. Alternatively, your learning style may identify you as being a team-player who thrives on group situations and projects and learns better when working in a team environment. The other benefit of knowing your learning style is that you can develop your strengths and work on any areas of weakness that fall outside your learning style to make you a more rounded individual and learner.

Learning styles and team working

Teams usually consist of individuals with different characters and different learning styles. There may be individuals within a group who are good at listening or making decisions or taking charge and leading the group. Without an awareness of learning styles, individuals could be misinterpreted; for instance, an individual showing initiative to lead and direct the group may be construed as being bossy.

By understanding learning styles collectively within a group it is easier to assign tasks and responsibilities to people who want to undertake certain roles because of who they are. If an individual is good at making decisions or organizing activities then you can play to their strengths. If an individual would rather sit back and listen and take notes, then this can be acknowledged and their needs and also the needs of the group can be accommodated.

Questions and review

1. There are a number of attributes that employers value from employees. These attributes can be classified into two areas – can you identify what they are?
2. Do you think it is better as an employer to employ people with a range of these skills or to focus on a specific skill or knowledge area only?
3. Why do you think that it is important to be able to work as part of a team?
4. What in your opinion are positive attitudes that should be promoted within the workplace?
5. Identify ways in which you could make communication more effective.
6. What techniques could you use to engage an audience?
7. How important do you consider interpersonal skills to be? Why do you say this?
8. Can you identify at least four barriers to communication and ways in which you could overcome these barriers to make communication more effective?

9. Written communication is very important because it can provide an auditable record of information exchanged between people. Written communication can also act as a guarantee and it can be legally binding, such as in a contract. For the following scenarios identify what key elements/criteria would need to be included within the document:
   - Producing a letter of complaint to a store manager regarding the purchase of a faulty item
   - Designing a CV to send off to an employer
   - Sending an email to the department informing them of a staff meeting

10. ICT can be used to support communication channels and make communication more effective. In what ways do you think ICT can support communication?

11. There are a number of specialist ICT channels that can be used to make communication more effective and dynamic. Some of these include: blogs, vlogs, podcasts and video conferencing. Can you provide examples of how and possibly where these could be used effectively?

12. There are a number of software and software tools available. Can you match three pieces of software to appropriate software tools? For example, the use of spell-check in a word processing piece of software.

13. A personal development plan (PDP) is one way in which you can identify your learning needs, record your objectives and review your achievements. Can you identify four ways in which you could put your PDP to use? Ensure that each way clearly identifies a need and has a range of targets.

14. Can you identify three ways of addressing PDP needs?

15. Complete a learning style questionnaire to identify your preferred learning style. What does your learning style tell you and how do you think you can benefit from knowing your learning style?

Assessment activities

Grading criteria	Content	Suggested activity
Pass		
P1	Explain why employers value particular employee attributes.	Research various job adverts and job roles and identify what attributes they are looking for in an employee OR produce a survey and ask employers what attributes they value in their employees. All information can be presented back to the group in a presentation.
P2	Describe potential barriers to effective communication.	Produce a leaflet that describes potential barriers to effective communication, for example: language, cultural, technical etc.
P3	Demonstrate effective communication-related interpersonal skills.	Give a presentation and produce a set of accompanying notes on a particular topic that demonstrates the use of a range of communication-related interpersonal skills. This could include the use of verbal exchanges, techniques and cues, summarising and paraphrasing and being actively aware of your environment and barriers that can impact upon effective delivery. You should also invite questioning at the end.
P4	Effectively communicate in writing technical information to a specified audience.	Produce a report aimed at the Director of Finance justifying the IT budget for the department based on the need to upgrade the company network and to purchase 50 more PCs across the branches.
P5	Use IT tools effectively to aid communications.	P5 can be integrated with other evidence criteria that demonstrate the effective use of IT to aid communications.
P6	Describe ways of identifying and meeting development needs.	In small groups identify ways in which development needs can be met. Produce a newsletter based on the group discussion, identifying ways to meet development needs.
P7	Create and maintain a personal development plan.	Create a PDP for a certain period (one month, a semester, a year). You will also need to demonstrate that you continue to review and update your target dates and objectives.
Merit		
M1	Explain and use a specialist IT communication channel effectively.	Explain and demonstrate to the group that you can use a range of specialist channels such as web pages, blogs, vlogs and video conferencing effectively.
M2	Explain mechanisms that can be used to reduce the impact of communication barriers.	In conjunction with P2, develop this further to identify mechanisms that can be used to reduce the impact of communication barriers.
M3	Proof read, review and amend both own and other people's draft documents to produce final versions.	In conjunction with P4, review a draft version of the report produced and make comments that can then be used to produce a final report. Review at least one other report from someone in your group and implement changes to your own report as a result of the review.
M4	Explain how individuals can use knowledge of their learning style to improve their effectiveness in developing new skills or understanding.	Participate in a learning styles activity to identify what type of learner you are, or your learning style. From this initial exercise you can then identify what your weaknesses are and state in your PDP (P7) how you will use this information in the future to develop new skills and understanding.
Distinction		
D1	Analyse interpersonal and written communications criticising or justifying particular techniques used.	Produce an information leaflet (at least two A4 sides) that analyses a range of interpersonal and written communications criticising or justifying particular techniques used. Base this justification/criticism on your own experiences of use.
D2	Independently use a personal development plan to undergo a process of identification of skill need and related improvement.	Identify new skills required to meet a given goal or target and discuss within your PDP how you are going to address this and bring about change and improvements over a set period.

Courtesy of iStockphoto, dlewis33, Image# 4473509

Computer systems are crucial in supporting a large majority of day-to-day organisational activities. Computer systems are quite complex in that they are based on a framework of integrated components that work together to provide a dynamic communications environment. These components can be broken down into hardware, software and peripheral items that individually have a set function such as storage, memory or speed etc. Together however, these individual components provide an integrated approach to addressing the growing business and communication needs of a multitude of end users.

Chapter 2

Computer Systems

The requirement to understand how computers work and how they interact and communicate with hardware and software components, and also the ability to diagnose, configure, install and provide routine maintenance is ever growing in today's competitive IT society. The ability to demonstrate knowledge and skill in these areas is desirable to employers and as such has generated the need for these topics to be embedded within qualifications at all levels of learning.

This chapter will introduce you to a range of concepts that will help you to develop an understanding about hardware and software components and considerations for use and selection. In addition, this chapter will provide information and guidance on how to carry out routine computer maintenance.

Each section in this chapter will focus on one of three learning outcomes:

- Understand the hardware components of computer systems.
- Understand the software components of computer systems.
- Be able to undertake routine computer maintenance.

Embedded within each section is a range of activities that will strengthen your understanding of the subject matter and provide you with the support you need to complete effectively some of the evidence requirements for the unit.

Understand the hardware components of computer systems

Computers are made up of a number of components that all work together like an interactive puzzle to ensure end-user processing and functionality in a range of task-based activities. In terms of hardware components, for each component there are numerous options with regard to speed, cost, capacity, storage and power capabilities; therefore, the selection of components should be based on meeting the needs and requirements of the end-user and the objectives of the organization.

System unit components

The central processing unit (CPU) is the main chip or processor that connects onto the motherboard in a computer. It is the brain of the computer. The CPU processes data that it fetches or receives from other sources. When people talk about CPUs, in the past they would refer to the clock speed, such as 3.2 GHz. With advances in technology, however, CPUs can be measured against their performance rating. Currently, two mainstream manufacturers make the CPUs used in personal computers (PCs): AMD and Intel.

The CPU is made up of different parts:

- the arithmetic and logic unit (ALU), which performs calculations and comparisons
- the control unit (CU), which controls the rest of the computer hardware
- the registers part (REG), which acts as a temporary storage area within the processor.

The main processing unit consists of a case, either a tower which is floor standing or a desktop unit on which a monitor would usually rest. The case protects the internal components such as the motherboard and acts as the interface between various input and output devices.

The motherboard (Figure 2.1) is the main circuitboard, which provides the base into which other hardware devices are plugged. The printed circuits on a motherboard provide the electrical connections between all of the devices that are plugged into it.

Devices that plug into a motherboard include:

- CPU
- basic input/output system (BIOS) memory
- random access memory (RAM).

In addition, there are slots that can be used for expansion boards and cards to provide the system with extra features. Expansion cards can be used for modems, graphics cards, TV cards, network cards, sound cards, small computer systems interface (SCSI) cards, etc.

Controllers are devices that transfer data from a computer to a peripheral device and vice versa. For example, visual display units (VDUs), printers and disk drives all require controllers. Controllers are often single chips that are preinstalled in most PCs. If, however, additional devices

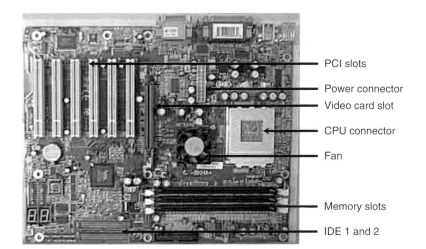

Figure 2.1 A motherboard
http://laray.com/build/board.html

are required these may have to be inserted separately via an expansion board.

To allow communication between different parts of the computer system, a 'bus' system is required. A bus is a group of parallel wires along which data can flow. The system bus is made up of a number of such communication channels that connect the processor and other components such as memory, input and output devices together. A computer will normally have several buses that are used for specific purposes.

In addition to these components, other features exist within the main processing unit, one of these being a coprocessor. The function of a coprocessor is to speed up certain types of computations; for example, a graphics accelerator card will sometimes have its own built-in processor to deal with graphics computations independently of the main motherboard processor.

The hard drive is an essential systems component, its primary function being to store the operating system and data including documents, graphics, games, utilities and software.

When physically installing a hard disk the power and data cables need to be connected and the master/slave jumper needs to be set as required if the hard disk is IDE (see Table 2.1). If another interface is used the master/slave configuration is not required. It is essential when a hard disk is being installed to tell the system the type and size of the actual disk in order for it to be detected and function accordingly. To do this, the BIOS setup must be entered in the system's startup sequence and the option to detect the newly installed disk should be selected.

A range of connectors, plugs and sockets is available in a computer system, a large majority of which is associated with the motherboard. On the motherboard there can be a number of ISA and PCI slots for expansion boards. Other sockets include ZIF sockets, IDE connectors for connecting hard drives, CD-ROM drives and backup drives such as ZIP drives, power supply connectors and port connectors (Table 2.1).

CHAPTER 2

Table 2.1 Plugs, sockets and connector types

Plug, socket or connector	What it does, what it connects to
ISA (industry standard architecture)	Plug and Play ISA enables the operating system to configure expansion boards automatically so that users do not need to fiddle with DIP switches or jumpers (which enable you to configure a circuit board for a particular type of computer or application)
PCI (peripheral component interconnect)	A local bus standard developed by Intel Corporation. Most modern PCs include a PCI bus in addition to a more general ISA expansion bus
ZIF (zero insertion force) socket	A chip socket that allows a user to insert and remove a chip without special tools
IDE (intelligent drive electronics or integrated drive electronics)	An IDE interface is an interface for mass storage devices, in which the controller is integrated into the disk or CD-ROM drive
BNC (bayonet Neill Concelman) connector	A type of connector used with coaxial cables. BNC connectors can be male or female and are used to connect cables and some monitors, which increases the accuracy of the signals sent from a video adapter. BNC connectors are used with high-frequency or noise-sensitive signals
Coaxial	A type of wire that has a grounded shield of braided wire that minimizes electrical and radiofrequency interference. Coaxial cabling is the primary type of cabling used by the cable television industry and is also widely used for computer networks such as Ethernet
RJ series plugs	Plug normally used for single-line (RJ-11) or multiline (RJ-12) telephone. The UTP (unshielded twisted pair) plug (RJ-45) can also be used for an Ethernet network line
DB series	A variety of male and female DB connectors, also called D-sub connectors. They normally come in 9, 15, 25, 37 and 50 pins or sockets

Parallel and serial ports

A parallel port is an interface for connecting an external device such as a printer to a computer. Most computers have both a parallel port and at least one serial port.

A parallel port uses a 25-pin connector to connect with printers and other devices requiring a relatively high bandwidth.

A serial port is a general-purpose interface that can be used for almost any type of device, including modems, mice and printers (although most printers are connected to a parallel port).

Peripheral items

Printers

Printers are categorized according to the technology that they use:

- **Line printers** – contain a chain of characters or pins that print an entire line at a time. Line printers are very fast, but produce low-quality print.
- **Dot matrix printers** – create characters by striking pins against an ink ribbon. Each pin makes a dot and combinations of dots form characters and illustrations.
- **Inkjet/bubblejet printers** – spray ink at a sheet of paper. Inkjet printers produce high-quality text and graphics.
- **Laser printers** – produce very high-quality text and graphics.
- **LCD and LED printers** – use liquid crystals or light-emitting diodes rather than a laser to produce an image on the drum.

When information is sent to a printer from the computer, the data is stored temporarily in a buffer. By buffering the data prior to processing, the application that has sent the data can be released to carry out other tasks.

Plotters

These fall under the category of 'large-format printer'. They are usually used for design work, and can manage poster images easily and professionally.

Plotters differ from printers in that they draw lines using a pen. As a result, they can produce continuous lines, whereas printers can only simulate lines by printing a closely spaced series of dots. In general, plotters are considerably more expensive than printers.

Memory

When software is loaded it must be stored somewhere inside the computer. The main storage space for programmes inside the computer is called 'memory'. Main memory is divided into a number of cells or locations, each of which has its own unique address.

There are two types of computer memory inside the computer: random access memory (RAM) and read-only memory (ROM).

RAM is the main store. This is the place where programmes and software, once loaded, are stored. When the CPU runs a programme,

Memory address – a number that is assigned to each byte in a computer's memory that the CPU uses to track where data and instructions are stored. Each byte is assigned a memory address. The computer's CPU uses the address bus to communicate which memory address it wants to access, and the memory controller reads the address and then puts the data stored in that memory address back onto the address bus for the CPU to use.

it fetches the programme instructions from RAM and executes them. RAM can have instructions read from it by the CPU and can also have numbers or other computer data written to it by the CPU.

RAM is volatile, which means that the main memory can be destroyed, either by being overwritten as new data is entered for processing or when the machine is switched off. Therefore it is not practical to store data files and programmes permanently in the main memory.

There are two types of RAM: static (SRAM) and dynamic (DRAM).

SRAM is incredibly fast and incredibly expensive. It is used as cache. Cache is a portion of memory made of high-speed SRAM instead of the slower and cheaper DRAM, which is used for main memory. Memory caching is effective because most programmes access the same data or instructions over and over. By keeping as much of this information as possible in SRAM, the computer avoids accessing the slower DRAM.

DRAM is used in most PCs. DRAM chips are available in several forms, the most popular of which are:

- dual in-line package (**DIP**) – can be soldered directly onto the surface of the circuitboard, although a socket package can be used in place of soldering
- small outline J-lead (**SOJ**) and thin, small outline package (**TSOP**) – can be mounted directly onto the surface of the circuitboard.

ROM will allow the CPU to fetch or read instructions from it; however, ROM comes with instructions that are permanently stored inside it and these cannot be overwritten by the computer's CPU. ROM is used for storing special sets of instructions which the computer needs when it starts up.

When a computer is switched off, the contents of ROM are not erased.

The more memory (RAM) a computer has, the better it works (Table 2.2). To measure how much a computer's memory will store we need to think of memory as a series of little boxes, referred to as memory locations, each of which is able to store a piece of computer data. The more locations that a computer's memory has is the key to its overall size. Computer memory is measured in units of 'thousands of locations' and in units of 'millions of

Table 2.2 Bits and bytes

Bit	Smallest unit of measurement	Single binary digit 0 or 1
Byte	Made up of 8 bits, amount of space required to hold a single character	Value between 0 and 255
Kilobyte (Kb)	Equivalent to 1000 characters	Approximately 1000 bytes
Megabyte (Mb)	Equivalent to 1 million characters	Approximately 1000 kilobytes
Gigabyte (Gb)	Equivalent to 1 billion characters	Approximately 1000 megabytes
Terabyte (Tb)	Equivalent to 1 thousand billion characters	Approximately 1000 gigabytes

locations'. More recently, vaster storage capacities have become available (Table 2.3).

Buffers are temporary storage areas, built in RAM, which act as a holding area, enabling the CPU to manipulate data before transferring it to a device.

Table 2.3 Bigger bytes

Petabyte	10^15 1 000 000 000 000 000	Approximately 1000 terabytes
Exabyte	10^18 1 000 000 000 000 000 000	Approximately 1000 petabytes
Zettabyte	10^21 1 000 000 000 000 000 000 000	Approximately 1000 exabytes
Yottabyte	10^24 1 000 000 000 000 000 000 000 000	Approximately 1000 zettabytes

Specialized cards

Specialized cards include network, graphics, video and sound cards.

- **Network cards** – sometimes referred to as network interface cards (NICs). The network card plays a very important role in connecting the cable modem and the computer together. It will enable you to interface with a network through either wires or wireless technologies. The network card allows data to be transferred from one computer to another computer or device.
- **Graphics cards/video cards and sound cards** – output graphics, video or audio data/information. Without a graphics card a computer will not function, unless there is an onboard integrated card.
- **Sound cards** – enable the computer to output sound through speakers connected to the motherboard, to record sound input from a microphone connected to the computer and to manipulate sound stored on a disk. A computer will function without a sound card as computers have integrated speakers, although these have very crude sound output. Nearly all sound cards support musical instrument digital interface (MIDI), a standard for representing music electronically. In addition, most sound cards are Sound Blaster compatible, so that they can process commands that have been written for a Sound Blaster card, the de facto standard for PC sound.

Activity 2.1

Hardware components are very interesting to look at and examine.

1. Examine a motherboard, looking at all of the circuitry and slots. Try to identify where the CPU would slot in if it is not already attached.
2. If you get the opportunity, examine a range of other hardware components and see how they all fit together in a case.

Backing store

There are many different types of backing store device. Some of these are internal, others are more portable, and each type has its merits and drawbacks in terms of storing and transferring data.

Disk drives and optical drives

Disk drives can be either internal or external to a computer; they read data from and write data to a disk. There are different types of disk drives for different types of disks. For example, a hard disk drive (HDD) reads and writes hard disks, a floppy disk drive (FDD) accesses floppy disks, a magnetic disk drive reads magnetic disks and an optical drive reads optical disks.

Hard disks were invented in the 1950s. They started as large disks of up to 20 inches in diameter holding just a few megabytes of data. These were originally called 'fixed disks' or 'Winchesters'. They later became known as hard disks to distinguish them from floppy disks. The performance of a hard disk can be measured in two ways, by the data rate and seek time. The other important parameter is the capacity of the drive, which is the number of bytes it can hold.

The storage capacity of an optical disk is vastly superior to that of other portable magnetic media, such as floppy disks. Data is read by and written to optical disks by lasers. The types of optical disks available include the compact disk – read-only memory (CD-ROM), write once read many (WORM) and 'erasable'.

Data can also be backed up to more portable media devices such as pen drives and flash memory cards, these devices providing excellent storage capacity.

> **Data rate** – the number of bytes per second that the drive can deliver to the CPU. Rates between 5 and 40 Mb per second are common.
>
> **Seek time** – the amount of time between when the CPU requests a file and when the first byte of the file is sent to the CPU. Times between 10 and 20 milliseconds are common.

Activity 2.2

Backing store comparisons

Complete the table below to provide valid and accurate comparisons between the different types of backing store.

Backing store type	Storage capabilities	Price range	Benefits	Limitations
Disk drive				
Floppy disk				
CD				
DVD				
Flash memory card				
USB pen				

Data transmission

There are various ways in which data can be transmitted internally within a computer, from one computer to another and between different electronic devices. To facilitate communicate paths and data transmission some sort of media or communication tool is required.

Data communication and transmission can be generated internally within a system or externally with the use of some sort of plug-in device.

Communication paths can be established with components such as buses and modems. In addition, other forms of devices such as Bluetooth, controllers and other wireless technology can facilitate transmission.

Buses

Buses are an example of how communication pathways are established within a computer. A bus is a collection of wires that facilitates data transmission between other components within a computer.

Buses consist of two parts:

- memory bus
- data bus

The memory bus identifies various locations within the main memory of a computer, whereas the data bus transfers actual information.

Modems

Modems are the main component for enabling data transmission and communication. They act as an interface and data converter from signals received from telephone lines, etc., to a computer, converting analogue signals into digital information (Figure 2.2).

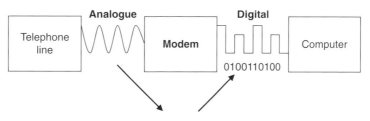

Figure 2.2 The role of a modem

Other devices

SCSI is a parallel interface standard that is used by a host of computer systems including Apple Macs, PCs and many UNIX systems for attaching peripheral devices to them. Unlike a parallel or serial port, SCSI has a faster data transmission rate (up to 80 Mb per second) than standard. In addition, you can attach many devices to a single SCSI port.

Bluetooth and Wi-Fi are examples of communication pathways that use radio signals to send and receive transmissions across devices such as mobile phones, headsets, personal digital assistants (PDAs) or computers.

Data transmission can also be influenced by other factors and devices such as the speed of a processor within a computer and the RAM speed, both of which can impact on the transmission media used.

Analogue – analogue transmission is used to carry voice, data or fax and has a limited bandwidth. Using a waveform, messages are carried by a cable such as a telephone line. Analogue information is transmitted by modulating a continuous transmission signal. A modem, for example, modulates data that is received over a telephone line in analogue to a digital format that can be understood by a computer, and demodulates received signals to retrieve data.

Digital – a digital signal is not a continuous wave like analogue. It consists of binary data sent in '1s' and '0s', 'on' or 'off', that is interpreted by computers. The Internet is a network of digital signals, as are most mobile phone technologies.

CHAPTER 2

The processor speed will impact on data transmission. In some cases you can override the original settings and 'overclock' the processor to make it run more quickly; however, this could cause the system to crash or, even worse, cause processor failure and meltdown.

The RAM speed can also impact on data transmission: the faster data can be read the faster it can be processed. Reading data from the memory at a faster speed will also ensure faster processing overall.

The impact of the actual transmission media, in terms of speed, connection, stability and reliability also needs to be taken into consideration; for instance as the physical connection, e.g. the cabling or wireless connection.

Activity 2.3

Carry out research into analogue and digital technology

1. Using three examples, such as modems, mobile phones and radio, identify what the difference in transmission is, and the benefits and limitations of each method (where appropriate).
2. Extend the table and include another option of your choice.

Technologies	Analogue	Digital
e.g. Modem		
Mobile phone		
Radio		

Considerations for selection

There are several factors and issues to consider when selecting hardware components for a computer. Some of these considerations are user based or cost based, while others are more functionally based.

When selecting hardware components it is crucial that you take into consideration the end-user and their requirements: what do they want from the computer?

Hardware components vary in terms of performance, capacity, functionality and cost. Depending on the environment and the end-user, a computer can be purchased for a few hundred pounds or thousands of pounds. Components can be quite simple and less expensive to suit the needs of a home user, or more sophisticated and more expensive to suit the needs of an organization. A home user may require a single computer, whereas an organization may require multiple machines, which will be a more expensive option.

End-users should always be considered when selecting components for a computer system. Different end-users have different needs and requirements, some of which could be classed as specialist, especially if the user has some sort of physical or mental disability that could restrict their interaction with a computer. Considerations include:

- How the user will interact with the computer – will they need special input or output devices?
- What sort of software is required?
- What outputs and peripheral items are required, such as printers, scanners and digital cameras?

Linked into user requirements is the issue of functionality. How do you want the system to perform, at what level, and are the user needs being addressed? For this, consideration should be given to the maintenance of the system, security, processing speed, power, storage capacity and compatibility with other systems.

Activity 2.4

A vast combination of hardware and software configurations can be used to build a computer system. The combination of hardware and software will be influenced by the need of the user and user type, for example home or business user. Cost will also be a crucial factor, as cost determines the speed, power, amount of storage and other key elements that can impact upon the task requirement and functionality.

Activity 2.5

For the following users, recommend a suitable system that will meet their needs. Include a full breakdown of hardware and software specifications and costs.

1. Jonathon would like a system on which he can carry out day-to-day general tasks at home, e.g. word-processing documents and using a spreadsheet for his household expenses. Jonathon is also a very enthusiastic player of games, especially simulation games, and requires a very fast computer with excellent graphics that can accommodate his needs.
2. RCW Trading is a small company based near Norwich. The company provides a range of imported specialist merchandise aimed at keen role-playing game (RPG) communities. The company has also established a strong following in terms of online forums, discussion groups and support for certain RPG online games, a service to which over 3500 users subscribe every month. The owner has two support people working for him based at his company branch. The requirements of the company are to have even more of an online presence. The projection is that within four months the subscription service should increase to 5000 users all requiring twenty-four-hour support. It has been decided that new systems are required to support this growing provision.

Understand the software components of computer systems

There are many different types of software. Operating system software is designed to support and enable the user to operate the system. Applications software provides the programmes and tools to design, model, input, manipulate and output data. This type of software includes spreadsheets, word processing, databases, graphics, presentation, communication and desktop publishing.

Applications software allows users to carry out information-processing activities, such as:

- word processing
- numerical and financial modelling and statistical analysis
- desktop publishing
- document management, presentation, storage, retrieval and manipulation.

Applications software also enables users to interact with hardware through different text, sound, animation, video and communicative techniques.

Communications software enables users to interact across different computers, platforms and networks, enabling the fast, efficient and secure transfer of data and information. Collaboration software provides the opportunity for groups of users to interact within a secure environment. Collaboration could take the form of an online discussion, verbal communication via a microphone or active participation in the creation or editing of documentation.

Utility software

This range of software provides the tools to support the operation, servicing and management of the system. Utility software is solution based, addressing issues such as security and system protection; for example:

- printer drivers
- virus checkers
- security software, e.g. firewalls
- defragmentation software
- partitioning software
- disk management software
- CD authoring
- DVD playback software.

Operating system software

Operating system software interacts with the hardware of the computer to ensure that the system resources are managed, controlled and coordinated. Operating system software is used on both standalone and networked systems and can be described as being text or graphically based. An example of a text-based operating system is MS-DOS, and graphically based systems include Windows 2000 and XP, and Mac X. Other operating system software also exists, such as Linux (Figure 2.3).

Linux is a free, UNIX-based operating system. Linux is popular in organizations because it offers good performance at relatively low cost. The only costs are associated with support and training, as the 'kernal' – the actual core operating system – is 'open source'. Linux is also

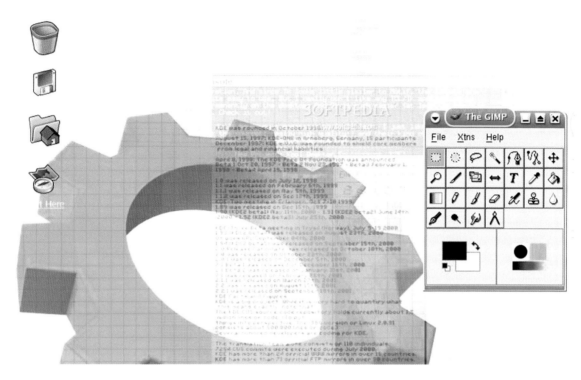

Figure 2.3 Linux screenshot

growing in popularity for mini-notebooks because they offer a lower specification for a reduced price and greater portability.

Open source – the actual code of any programme is freely available to modify.

Another type of operating system is Mac OS and more recently Mac OS X (Figure 2.4), which is graphical user interface (GUI) based and specifically used with Macintosh systems. Mac OS X is built around UNIX, so it is reliable and quite easy to operate. In addition, its business applications and communications are as fast and functional as a Windows-based system.

The Windows OS environment (Figure 2.5) is once again GUI based. It has evolved over the years into one of the biggest, if not the biggest, home-based user operating systems.

Command line and graphical user interface operating systems

Command line systems include UNIX and MS-DOS (Figure 2.6), where the operating system is text based as opposed to graphically based.

GUIs, such as Microsoft Windows XP and Vista, can be characterized by a number of features, including:

- **A pointer** – a symbol that appears on the screen that allows you to move and select objects or commands. The pointer usually appears as a small angled arrow.
- **A pointing device** – such as a mouse or trackball, that enables you to select objects on the screen.
- **Menus** – most GUIs allow the user to carry out tasks and functions by selecting an option from a menu (Figure 2.7).
- **Icons** – these are miniature pictures that represent commands, files or windows (Figure 2.8). The pictures represent pictures of the actual

Figure 2.4 MAC OS X Screenshot

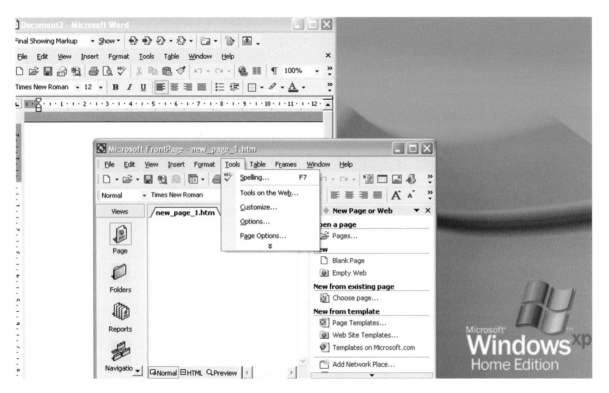

Figure 2.5 Windows Screenshot

Figure 2.6 MS DOS

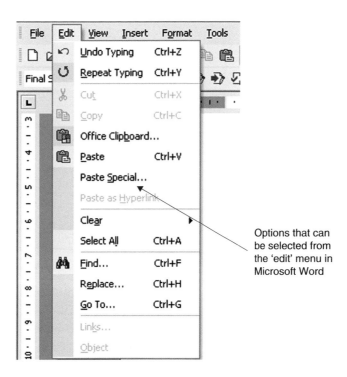

Figure 2.7 Example of a GUI menu option

commands, for example a rubbish bin to represent 'rubbish' and a disk to represent 'save'. When the picture is selected the command is executed.

- **Desktop** – this is the area on the display screen where icons are grouped (Figure 2.9). This is often referred to as the desktop because the icons are intended to represent real objects on a real desktop.
- **Windows** – these can divide the screen into different areas. In each window, a different programme or file can be displayed.

Figure 2.8 Example of icons

Figure 2.9 Desktop icons

Operating system functions and services

Computers need operating system software to function. Operating systems perform basic tasks, such as recognizing input from the keyboard, sending output to the monitor or display, keeping track of files and directories on the disk, and controlling peripheral devices such as printers and scanners. In large corporate systems the operating system has an even greater responsibility in that it acts as a mediator to ensure that programmes running simultaneously do not interfere with each other. The operating system also ensures that users do not interfere with the system, by restricting access to certain areas.

The main functions of an operating system can be classified under machine and peripheral management, and security and file management.

Tasks that are typically performed under machine and peripheral management include:

- setting the time and date, printer, mouse and keyboard configurations
- scheduled tasks

- GUI desktop and display setup
- power management
- applications software icons.

Machine and peripheral management can be carried out through the control panel on the operating system, although some, such as the setting of the time and date, can also be done through the ROM-BIOS startup.

The control panel (Figure 2.10) illustrates how the majority of settings can be adjusted to suit user needs. These include the time and date (Figure 2.11), printer, mouse and keyboard configurations (Figure 2.12)

CHAPTER 2

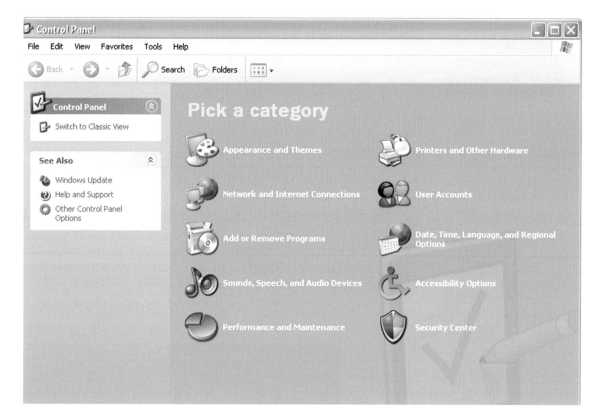

Figure 2.10 Control panel features

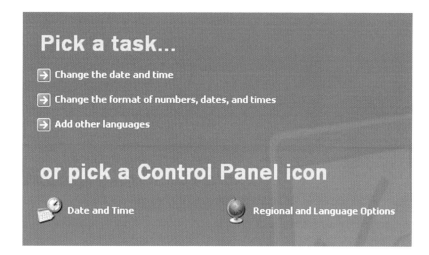

Figure 2.11 Time and date settings

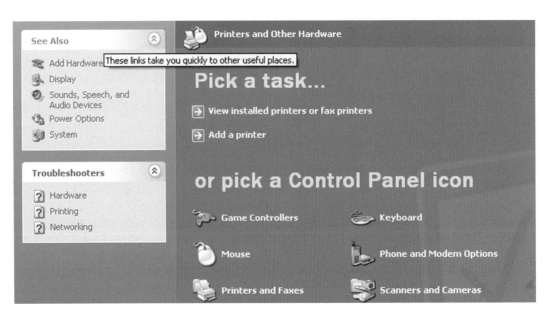

Figure 2.12 Printer configurations

and password properties. In addition, user accounts can be set up to protect and restrict log-ins to certain applications.

Scheduled tasks such as backing up data can be carried out through the operating system (Figure 2.13). Depending on whether you are using a networked or standalone computer, the backup procedure will download your data onto a local hard disk or onto the network, where specific storage areas are designated for individual users. With most graphical operating systems, procedures for carrying out scheduled tasks involve

Figure 2.13 Scheduled tasks

either 'point and click' or 'drag and drop'. On the performance and maintenance menu in the control panel you can also carry out a range of scheduled tasks.

Through the operating system you can change the way your GUI desktop and display setup looks. This can be done through the 'appearance and themes' setting on the control panel (Figure 2.14).

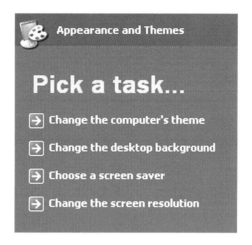

Figure 2.14 GUI appearance

Security functions include setting up and assigning password properties or virus protection configuration. File management functions include directory (folder) structure and settings.

Software utilities

Utility software can incorporate a range of features and functions, such as virus protection, firewalls, cleanup tools, e.g. cookies, Internet history, defragmentation and drive formatting.

Activity 2.6

1. Select three different types of utility software and explain their functions.
2. Different virus protection software packages are available on the market:
    - Identify three different pieces of virus protection software.
    - Identify the manufacturer, cost and some of the functions and features.
    - Select which piece of virus protection software you feel is the best value for a home user.

Be able to undertake routine computer maintenance

Computers require maintaining, just like cars or any other system appliance. Although you may be a regular user of a computer, you may not necessarily know how to maintain it. A novice user could undertake routine operations such as upgrading software; however, more complex software and especially hardware maintenance may require the input of a specialist.

Computer maintenance can be divided into three parts: software maintenance, hardware maintenance and general file management.

Software maintenance

Software maintenance can cover a wide range of tasks and there are several factors to take into consideration (Figure 2.15). Some of these

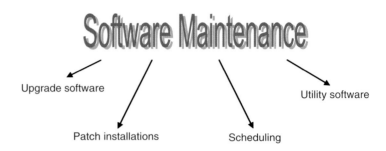

Figure 2.15 Software maintenance considerations

tasks can be classed as routine and may be carried out on a daily or weekly basis, whereas others may only require periodic maintenance, for example once every month or three months. Some software maintenance tasks will be on demand, some as and when required.

Upgrading software ensures that any new tools and features are available to the end-user. For example, a new piece of applications software or a newer version may have better formatting or reporting features. Software upgrades can also be driven by software manufacturers no longer offering support for older versions. This is especially apparent with operating system software. Upgraded software may:

- have greater productivity and efficiency
- be faster
- be more reliable and dynamic
- have more features
- have a new interface.

Patch installations are updates that usually fix a bug or software error. Patches can also be used to support new hardware in terms of resolving communication and conflict issues between systems or platforms or providing driver solutions to components such as graphics cards.

Scheduling of maintenance tasks is important, especially in terms of trying to complete installations and upgrades at non-critical times when end-users do not require access to the system. Some maintenance tasks may have to be performed during the day, especially if they are critical to the system, such as security maintenance.

Routine computer maintenance tasks include updating, installing and working with a range of utility software. Certain utility software can be used generically across any system to clean up or provide protection against viruses or other security breaches. Some users, however, may require specific functions from utility software such as defragmenting a hard disk to rearrange files more efficiently to allow for faster data access (Figure 2.16), or compression facilities to reduce the size of files to increase the available space on a hard disk.

Figure 2.16 Disk defragmentation option

Hardware maintenance

Hardware maintenance can include a range of routine and non-routine tasks, such as:

- cleaning equipment
- installing or configuring new additional or replacement devices
- compliance with regulatory requirements such as health and safety.

Cleaning equipment

Ensuring that equipment and components are kept clean and dust free can contribute to their longevity. Some organizations conduct cleaning on a periodic basis, others clean as and when users require it.

Installing and configuring devices

Installing and configuring new, additional or replacement components can be classed as a routine task, especially with 'plug and play' technologies that require minimum configuration or user expertise.

Some initial steps and precautions should be taken when installing hardware:

- shutting down and switching off your computer
- unplugging the power cord from the wall socket or rear of the computer.
- giving yourself sufficient room to move around the computer and desk area.

Computer chips and hardware such as motherboards and hardware cards are sensitive to static electricity. Before handling any hardware or working on the inside of the computer you should ensure that you have discharged the static electricity from your body. At the very least you should unplug the computer from the mains and touch the bare metal case with your hand to discharge any static that may have built up in your body. Ideally, a grounding strap should be worn and an antistatic mat used to reduce the risk of any components being 'zapped' by static.

The majority of computers that are bought today have preinstalled applications software as part of the package. If, however, additional software needs to be installed this can be done easily by inserting the disk, which then launches a step-by-step visual installation process.

If you are installing a new hard drive you should put it in a closed front bay (one that cannot be accessed from the front of the PC case).

If you are installing a new CD or DVD-ROM you will need to put it in an open bay which can be accessed from the front of the PC case. An open front bay is easy to locate because of the removable front panel which can be unclipped once the case is open. There may also be a metal trim which has to be snapped off before inserting the new drive.

Installing a mouse used to involve physical connection and then configuration so that the mouse and the computer could communicate with each other. These days, however, especially with a universal serial bus (USB) mouse, the only real requirement is to connect the mouse physically.

When installing expansion cards, you need to identify the type of slot to install your new hardware into. If there is no spare slot of the type you need you will have to remove one of the other cards in your computer to use the new one.

If you are installing the new hardware into a previously unused slot you will probably have to remove the backing cover from the case before you proceed. This is a metal clip or cover designed to stop dust, etc., from getting inside the case. They are normally held in place by a single screw or can be 'snapped' out of the casing using pliers. Remove the retaining screw and/or unclip the backing cover.

Printer installation and configuration involves a physical connection using a parallel, serial or USB cable. In terms of software, most printers have accompanied printer drivers that can be installed to ensure that the computer and the printer are communicating on the same level. In most modern-day operating systems, however, printer configurations can be selected from a list covering the majority of printer makes.

Regulatory requirements

Compliance with regulatory requirements can impact on the need to conduct system audits and additional hardware maintenance. Compliance with health and safety regulations may require the movement of systems and wiring, the installation of more up-to-date or robust components, or even a move towards a networked or wireless working environment.

The working environment of users of ICT is of major importance when it comes to issues such as health and safety. Relevant points cover environmental, social and practical aspects of working conditions.

Users should be working in an environment that has adequate ventilation and natural lighting, and the temperature should be conducive to a computing environment, especially as computers give out large quantities of heat.

Computers should also have sufficient support peripherals such as filter screens to minimize glare and height-adjustable chairs. When working at a computer no food or drink should be consumed, in case liquid or crumbs fall onto the keyboard or into the case. Wires should always be packed away in appropriate conduits and not left trailing across the floor.

Hardware maintenance could address many of these health and safety requirements. Systems may have to be moved to ensure that users have access to natural lighting and adequate ventilation. Additional trunking, ducts or conduits may have to be installed to ensure that no wires are hanging, or the organization may even move towards a wireless environment.

Some considerations are for the benefit of the user, such as breaks, adjustable chairs and filter screens. This could lead to the type of system that they use becoming more portable and less desk dependent.

Computer maintenance has to take into consideration laws and regulations. One health and safety Act, The Display Screen Equipment (VDU) Regulations 1992, was set up to protect users of information systems and general ICT. Under these regulations an employer has six main obligations to fulfil.

For each user and operator working in his undertaking, the employer must:

(i)   assess the risks arising from their use of display screen workstations and take steps to reduce any risks identified to the 'lowest extent reasonably practicable'

(ii) ensure that new workstations ('first put into service after 1st January 1993') meet minimum ergonomics standards set out in a schedule to the Regulations.
      Existing workstations have a further four years to meet the minimum requirements, provided that they are not posing a risk to their users.

(iii) inform users about the results of the assessments, the actions the employer is taking and the users' entitlements under the Regulations.
      For each user, whether working for him or another employer (but not each operator)

(iv) plan display screen work to provide regular breaks or changes of activity.

In addition, for his own employees who are users,

(v) offer eye tests before display screen use, at regular intervals and if they are experiencing visual problems. If the tests show that they are necessary and normal glasses cannot be used, then special glasses must be provided

(vi) provide appropriate health and safety training for users before display screen use or whenever the workstation is 'substantially modified'.

Such legislation does indeed benefit users and helps to protect them against computer-related injuries, such as:

- repetitive strain injury (RSI)
- back and upper joint problems
- eye strain
- exposure to radiation and hardware ozone
- epilepsy
- stress-related illnesses.

However, from an employer's point of view, maintenance of the hardware and updating components, peripheral items, cabling and other safety features can be a very expensive venture.

File management

File management collectively looks at the process of file housekeeping. This could entail editing, deleting, movement, storage and backup of files and folders.

Creating folders is very simple to do in the majority of operating systems. Folders allow you to store a number of files in a single location. For example, you could set up folders for each unit that you are studying on the National Diploma and each folder will contain files related to that unit, e.g. assessments, exercises, notes and draft copies of work.

Activity 2.7

1. Set up a folder for each of the units you are currently studying on your BTEC National for IT Practitioners.
2. Name all of your folders appropriately by the unit number and/or name.

Backup procedures should also be considered under file management. Backups can be generated in a number of ways, and these will differ for each type of user, for example a home user or an organization, and also in the number of systems and quantity of data. Other backup issues may include the nature or critical aspect of the data, as this can have an impact on the frequency of the backup and the type of component used to perform this function.

Backup options include:

- simple backup
- stack backup
- advanced stack backup

- incremental backup
- grandfather, father, son backup.

Simple backup is the elementary backup type. Each time an archive is created the oldest version of the backup file is replaced with the newly created one.

Stack backup consists of the last created backup and previous versions, the previous versions being organized into a stack format.

The advanced backup procedure differs in that it does not permit unchanged or unedited files in the old backup version copies to be stacked.

An incremental backup provides a method of backing up data that is much faster than a full backup. During an incremental backup only the files that have changed since the last full or incremental backup are included. As a result, the backup may be conducted in a fraction of the time it takes to perform a full backup.

The grandfather, father, son technique is probably the most common backup method, which uses a rotating set of backup disks or tapes so that three different versions of the same data are held at any one time. An example of this method is shown in Table 2.4.

Table 2.4 Grandfather, father, son backup method

Customer order data					
Monday		**Tuesday**		**Wednesday**	
Disk 1	Grandfather	Disk 2	Grandfather	Disk 3	Grandfather
Disk 2	Father	Disk 3	Father	Disk 1	Father
Disk 3	Son	Disk 1	Son	Disk 2	Son

For a home user, backing up data may simply involve saving it to a hard disk and another media format, such as a CD or USB pen.

References

The following texts should further enrich your knowledge and understanding of computer systems:

Englander, Irv (2003) *The Architecture of Computer Hardware and Systems Software: An Information Technology Approach*, 3rd edn., John Wiley and Sons.

Newman, Robert (2008) *Computer Systems Architecture*, Lexden Publishing.

Nisan, Noam (2008) *The Elements of Computing Systems*, MIT Press.

Shooman, Martin (2002) *Reliability of Computer Systems and Networking Fault Tolerance, Analysis and Design*, Wiley.

Williams, Rob (2006) *Computer Systems Architecture: A Networking Approach*, 2nd edn., Pearson/Prentice Hall.

Questions and review

1. Can you identify five components within a computer system and describe how they contribute to the functioning of the system?
2. What do IDE and EIDE mean?
3. What performance factors do you need to take into consideration in relation to backing store?
4. There are a number of factors that need to be taken into consideration when selecting a computer system. Can you identify three of these factors?
5. There are a range of different operating systems such as Windows, Linux and MAC OS. Can you produce a table that compares and contrasts each of these?
6. There are a range of different software utilities. Can you identify four different utilities and describe what they do/what function they perform?
7. You are expected to carry out a range of routine computer maintenance tasks. Demonstrate that you can perform the following tasks:
   - Upgrade software such as virus detection files
   - Install patches
   - Schedule maintenance tasks
8. Hardware maintenance tasks are required to replace or upgrade existing or faulty components. Demonstrate that you can install and configure a printer and a replacement device.
9. What health and safety issues would you need to consider when replacing hardware components?
10. What elements can be considered under 'file management'?

Assessment activities

Grading criteria	Content	Suggested activity
Pass		
P1	Explain the function of the system unit components and how they communicate.	Produce a technical guide that explains the functions of system unit components and how they communicate (you could use annotated illustrations to further enrich this evidence).
P2	Describe the purpose, features and functions of two different operating systems.	Show evidence of researching two different operating systems, for example Windows, Linux, Mac, Solaris, AIX etc. Produce a table to describe the purpose, features and functions of the two.
P3	Demonstrate the operation and explain the use of two different software utilities.	Demonstrate the operation of two different software utilities and write a short summary explaining what they do, their functions and features etc.
P4	Describe the range of available utility software.	Produce an article aimed at a technical forum submission that describes the range of available utility software.
P5	Undertake routine maintenance tasks in relation to a PC.	Demonstrate your competency in carrying out routine PC maintenance tasks. Use an observation sheet or witness statement as evidence in conjunction with accompanying screenshots.
Merit		
M1	Explain and implement the installation and configuration of an additional or replacement device.	Imagine a scenario where you have been asked to install and configure an additional or replacement drive. You have been asked to perform this practical task as part of an interview selection process to demonstrate your technical skills. To complement this, you will also need to explain the installation and configuration process – this can be in the form of an instruction guide.
M2	Compare the features and functions of two different operating systems.	Select two different operating systems and compare their functions and features. This is probably best approached by using a table to present the evidence.
M3	Explain the effect of the software maintenance activities carried out on the performance of a computer system.	Produce a report, possibly in conjunction with M1, following through the interview selection process to demonstrate your knowledge of the 'effects of software maintenance activities carried out on the performance of a computer system'.
Distinction		
D1	Evaluate at least three specifications for commercially available computer systems and justify the one most suitable for use in a given situation.	Using a scenario such as 'you have been asked by the owners of a small DVD chain of shops to help them in their decision making process of updating their existing computer system', research at least three specifications and justify in a report the most suitable system for the shops.
D2	Justify the considerations for selection in the upgrade of an existing computer system.	In conjunction with D1, add a section within the report justifying the considerations for selection in the upgrade of an existing computer system.

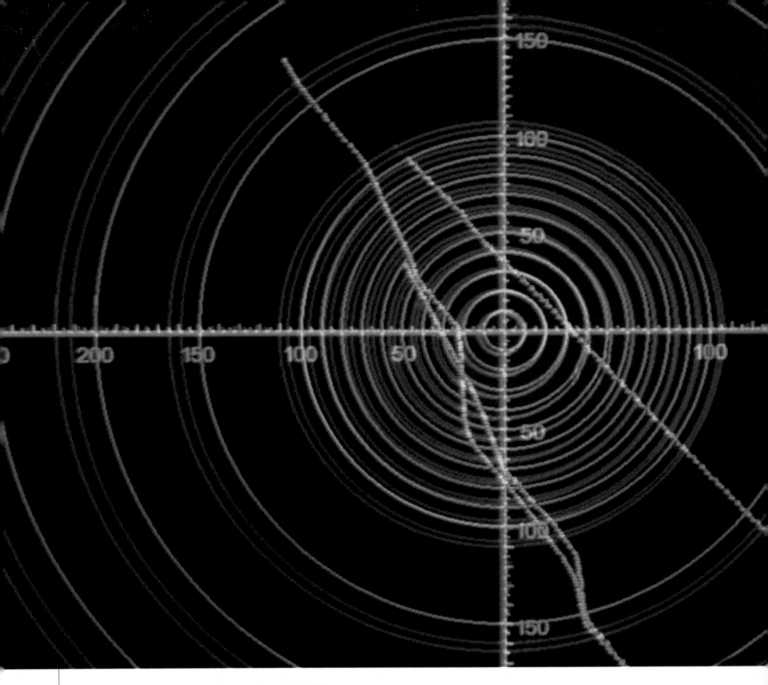

Courtesy of iStockphoto, luoman, Image# 4348856

Data and information and the communication of these are paramount to an organisation. The need for information that is current, accurate, timely, reliable, and consistent and of value will ensure that the right systems are put in place to enforce any business objectives and strategies.

Transmitting this information and communicating it both internally and externally is also key to the success of an organisation. This is why there are a range of information systems available to support users at all levels in this goal.

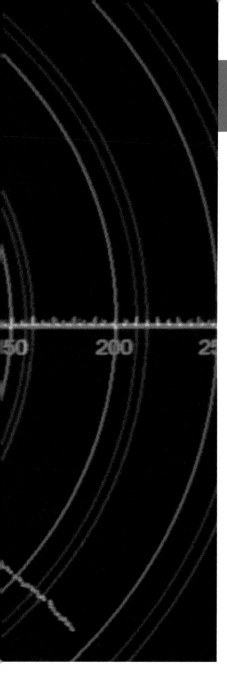

Chapter 3

Information Systems

Information is used in a variety of contexts to help plan and forecast for the future. Information is also used to help make decisions, problem solve and analyse situations. Information systems are a combination of hardware, software and expertise that is used to support all of these functions.

Information systems provide a mechanism for communication and also a tool that can be used to input, process and output information that can be used at all levels within an organisation.

This chapter will examine how data and information can be used in organisations and how different functional areas of an organisation can utilise this resource. In conjunction we will also examine the specific functions and features of information systems and the benefits and constraints of use.

The chapter will be structured around the following learning outcomes:

- Know the source and characteristics of business information.
- Understand how organisations use business information.
- Understand the issues and constraints in relation to the use of information in organisations.
- Know the features and functions of information systems.

Know the source and characteristics of business information

Business information is crucial to the success of any organization. Information can be derived from a number of different sources, it can be processed in a variety of different ways depending on the business needs and it can be output in various formats.

Characteristics of information

Data and information form the backbone of organizations. Data is the raw, unprocessed element that could consist of random, even meaningless code until it is converted into a meaningful format, when it reverts into information. Information is the processed product that can then drive a business into becoming more efficient, productive, competitive or cost-effective.

Information can be characterized in a number of different ways that reflect the overall quality of the information. O'Brien (1993) describes the attributes of information quality as being divided into three different categories:

- time
- content
- form.

For each category a number of attributes exists, as shown in Table 3.1.

Table 3.1 Attributes of information

Attributes that describe the quality of information			
Time	**Content**	**Form**	**Additional characteristics**
Timeliness	Accuracy	Clarity	Confidence in source
Currency	Relevance	Detail	Reliability
Frequency	Completeness	Order	Received by correct person
Time period	Conciseness	Presentation	Sent by correct channels
	Scope	Media	

The time category describes the period of the information and the frequency in which it is received. Further breakdown of the attributes identifies that the information should be available when needed (timeliness). The information should reflect what is going on in the current environment and be adaptable for changes in the future (currency). There is a need to have access to information time and time again at regular intervals, every minute, hour, day, week, etc. (frequency), and quality information that covers the correct period; for example, to check back over last year's sales, historical information would be required (time period).

The content category examines the correctness of the information supplied to an organization (accuracy), how applicable the information is to a given situation (relevance) and how comprehensive the information is (completeness). This category also looks at whether the information supplied is presented in an applicable format, e.g. graphs or charts used to show large quantities of numerical information such as sales figures (conciseness). Finally, the information supplied should be appropriate and meaningful to the recipient (scope).

The form category ensures that the information is presented in a clear and appropriate format for the recipient (clarity), and at the correct level and depth (detail).

Other quality factors in this category examine the ordering of the information and ensure that information is delivered correctly (order), how the information is presented and whether the format is clear (presentation), and whether the tool of transmission is appropriate (media), e.g. releasing sales figures on a report and not an e-mail.

Additional characteristics can be used to assess the quality of information; these include confidence in the source of the information and issues of reliability.

Other more generic characteristics of information exist beyond the quality aspect. These include:

- **source** – internal, external, primary, secondary
- **nature** – quantitative, qualitative, formal, informal
- **level** – strategic, tactical, operational
- **time** – historical, current, future
- **frequency** – real-time, hourly, daily, monthly
- **use** – planning, control, decision
- **form** – written, visual, aural, sensory
- **type** – disaggregated, aggregated, sampled, random.

Each of these characteristics provides a more holistic view of how complex information can be.

The source of information is important because it raises questions about how reliable the information is: if the information has been passed down through a number of channels (secondary information), how accurate is it? Internal and external information will affect the day-to-day operations of a business differently, as shown in 'Sources of information' below.

The nature of the information will affect how it is presented, interpreted and passed on. Quantitative information is based on facts and statistics, key information used for planning, forecasting and decision making, etc.

Examples of this type of information include monthly expenditure, sales figures and employee performance status. Quantitative information is essential if you are working with large data sets because facts and figures are easier to map and model than descriptive qualitative information.

CHAPTER 3

Qualitative information provides the details, giving additionality to any existing information. For example, you may know that a customer shops weekly, but the qualitative aspect will identify what they buy, when they buy it, trends in buying, and so on. One of the best ways to extract qualitative information is through interviewing. Interviewing allows you to obtain controlled qualitative information that can serve as a balance to pure facts and figures.

Within an organization there are three distinct levels, and information requirements at each level will be different. Using the example of a supermarket, a breakdown of information at each level is provided:

- **strategic** – plans to open/close branches, introduce new/scrap old product lines, invest more money into certain areas, e.g. bakery and fresh produce
- **tactical** – increase/reduce staffing levels, change or introduce new job roles, training and development, decide which product items to stock over a certain period and in what quantity
- **operational** – checking that shelves are stocked, reordering of stock, checking shelf-life of stock items, correct pricing of items, changing promotional material, checking that stock items are stacked/positioned correctly.

The characteristic of time allows users to identify when in a certain period information was generated and communicated, and how significant that information is to the current environment. Some information is deemed to be irrelevant because it is too outdated; for example, how much it would cost to buy a new computer now, based on last year's prices? However, historical data can be useful especially if an organization is looking back to identify trends over a certain period in order to make predictions for the future.

The frequency of information is very important within an IT environment and especially when information systems are being used. The frequency at which information is input into a system, processed and output could save an organization time and money. For example, up-to-date information about customer buying patterns is important.

Information can be used for many different purposes within an organization, including:

- planning
- processing
- forecasting
- decision making
- controlling
- supporting and guiding.

The use of an information system can help managers and submanagement levels to carry out these functions more efficiently, cost-effectively and productively.

Information can take on a number of different formats, including verbal, written, visual and expressive (Table 3.2). Throughout the course of a day any number of these ways of communicating information may be used.

Table 3.2 Information formats

Verbal	Written	Visual	Expressive
Directing	Letters	Charts	British Sign Language
Advising	Memos	Maps	Pointing
Informing	Minutes	Graphs	Smiling
Challenging	Reports	Pictures	Frowning
Debating	Agendas	Designs	Laughing
Persuading	Invoices	Moving images	Crying
Delegating	Statements	Drawings	Hugging
Enquiring	Receipts	Static images	Waving

The way in which information is presented in terms of its structure and type can be dependent on the initial data source. For example, information about how many consumers buy a certain product could be based on survey information that has been sampled. Other information types include aggregated, disaggregated and random.

Activity 3.1

Complete the following table by giving an example of how the piece of information could be used within an organization and by whom.

Information requirement	Example of use	By whom
Planning		
Processing		
Forecasting		
Decision making		
Controlling		
Supporting and guiding		

Sources of information

Organizations need information to enable them to function on a day-to-day basis. The information requirements of an organization can be broken down into two broad categories: internal information and external information.

Internal information can be derived from a range of functional departments such as financial, personnel, marketing, purchasing, sales, manufacturing or administration. External information can come from government sources or trade groups, or it could be provided commercially, for example by an organization that buys consumer data on buying patterns or consumer trends. Databases are a good source of external information, from which access to customer, patient, competitor or financial data can be retrieved. Information obtained through research has the benefit of being current, authentic and potentially more relevant to a niche market.

Whatever the source of information, internal or external, the most important factor is that the source is reliable.

Internal information

Internal information can be accessed from the following sources:

- employees
- functional departments (sales, IT, human resources, marketing, finance, operations, etc.)

The type of information that is generated and captured by departments includes the following.

- Sales
  - how many product or service types have been sold?
  - how quickly were these product or service types sold?
  - pricing strategies
  - the feasibility of launching a new product or service
  - whether to diversify into new markets or ranges
- IT
  - issues with the computer systems: crashes, downtimes, new systems online, etc.
  - number of users on the system
  - security
  - IT strategies: designing a company website, investing in upgrades, networking issues, etc.
- HR
  - training strategies
  - hiring and firing
  - disciplinary actions
  - internal policies and procedures, e.g. health and safety, code of conduct, equal opportunities
- Marketing
  - what product/service to promote
  - when to promote
  - how much to promote
  - where to promote and in what format
- Finance
  - profit, loss and budgeting
  - money to invest, diversify
  - deficit and growth areas
  - salaries and pensions
- Operations
  - how many products have been produced?
  - productivity levels
  - raw materials and cost of production
  - stock levels
  - delivery and dispatch issues.

External information

External information is information that has come from outside the boundaries of an organization. External information sources can be seen in Figure 3.1. Information requirements from these sources will influence the way in which an organization will function. For example, a supermarket case study and three of the more common external sources are presented in Table 3.3.

Figure 3.1 External information sources

Table 3.3 Information requirements from external sources

Captured from	Information requirement	Impact
Customers	What customers buy	Alter stock requirements to meet demands
	When customers buy	Put on promotional offers to tempt customers to buy more of a product or similar products
	How much they buy	Move the positioning of certain stock items
	How much they spend	Send out coupons and money-off vouchers for certain products
	Where they buy it	
Suppliers	How much they can supply	Certain stock items may only be offered at certain times, e.g. new potatoes only in the spring/summer months
	When they can supply it (seasonally)	Fluctuations in prices due to supply and demand
	How quickly they can deliver it	Use additional suppliers, to ensure delivery and quality or cut costs
	How much it will cost	
	Reliability	
	Quality of product	
Competitors	Market share	Cut costs, improve operations, reduce overheads, offer more products and expand the business
	Current offers and promotions	Increase promotional activity
	Pricing policies	Carry out market research: what do customers want?
	Items stocked	Increase/decrease prices
		Stock more or fewer products

Understand how organizations use business information

Business information is used to support companies in their everyday decision making, planning and organization. Business information serves a number of different purposes that can help an organization to become more competitive, productive or profitable. The information can be used by individual users, functional areas such as sales or IT, or senior managers and strategic decision makers.

One of the most important features of business information is how it is used and communicated throughout an organization. Information flow is critical to ensure that ideas and plans are distributed throughout each level of an organization, from the very top to the very bottom.

Purposes

Business information can be used for a number of different purposes, including:

- operational support
- analysis
- decision making
- gaining competitive advantage.

Operational support can cover a multitude of functional processing activities and tasks that are performed on a daily basis. Business information can help in terms of monitoring and controlling these activities. For example, when stock levels are running low, automatic reordering could take place or alerts could be provided to indicate that stock levels are diminishing.

Analysing patterns and trends is crucial to any organization, and business information can help to support these critical tasks. If, for example, your core business was retail, it would be useful to identify the buying patterns of your customers in terms of:

- age
- gender
- socioeconomic background
- average spend
- products purchased
- when purchased.

Analysis of information is very important to organizations, to the extent that data is captured at the point of sale with devices such as loyalty and reward cards (Figure 3.2). These cards provide information to the organization that then allows them to target their marketing more specifically. A good example of this would be sending a customer discount vouchers for products recently purchased or products of a similar nature.

Figure 3.2 Loyalty and reward cards: (a) WH Smith 'Club card'; (b) Tesco 'Clubcard'; (c) Sainsbury's 'Nectar'; (d) Boots 'Advantage Card'

Activity 3.2

1. **Produce a list of as many loyalty/reward card schemes as you can.**
2. **From the list, pick five cards and identify what they are offering in terms of rewards and incentives to the customer.**
3. **How many different loyalty cards do you or your family have?**

Decision making

Within an organization decisions are made on a daily basis. These decisions vary according to the level at which they are taken and the complexity of the decision to be made.

Within a typical organization three levels of management exist, each level representing a different decision type (Figure 3.3).

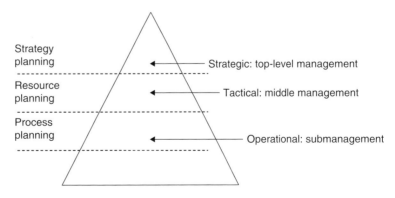

Figure 3.3 Levels and types of decision making

The strategic level represents the highest levels of management, i.e. managing director, chief executive and senior managers. Decisions at this level encompass planning for the future, strategy decisions such as mergers and takeovers, forecasting markets and trends.

The tactical level represents middle management levels, including heads of departments and assistant directors. The decisions focus on project plans, resource issues and financing.

The operational level focuses on day-to-day decision-making tasks, the mechanics of the organization. This level, which is positioned at the base of the organizational structure, includes the majority of the workforce at submanagement levels.

One of the other purposes of business information is to use it to gain a competitive advantage. If you have a breakdown of your products and services, customers and their buying patterns, predictions can be made as to the peaks and troughs throughout a certain period. Competitor information can identify how the market share is divided, what promotions and marketing campaigns are planned by competitors and possibly how effective they are or have been.

Functional areas

Most organizations, especially large organizations, are structured into departments or functional areas such as sales, purchasing, manufacturing, marketing, finance, personnel, administration and IT. Each functional area will use a range of business information to support the processing of activities carried out on a daily basis (see 'Sources of information' above).

Some business information that is used by the different functional areas is quite specialist; for example, the auditing of hardware and software may be data that is retained by an IT department, whereas sales performance figures may be required by the sales, marketing and finance departments.

Information flows

Information flow within an organization can be influenced by a number of factors, such as:

- size and structure
- information type/nature
- tool of delivery
- source and recipient.

The size and structure of an organization can have a profound effect on how easily and quickly information flows within it. Chapter 6 examined different types of organizational structure, flat and tall/hierarchical, with the relative merits and drawbacks of each.

In terms of information flow, tall organizations could promote a better information flow environment owing to the clear divisions into functional or specialist departments, e.g. sales, finance, insurance or claims. The information flow would be more direct and therefore the speed of communication could be that much quicker. In a smaller, flatter organization, information flowing into and out of the organization may be slower, as no one person would be directly accountable.

The type/nature of the information could also influence effective information flow within an organization. Formal information such as the passing of a customer order from sales to dispatch may be deemed to be of priority, resulting in a quicker flow of information. Internal communication such as the announcement of the next team briefing, may be passed on to departments with less urgency.

The tool used to deliver information can also affect the flow and speed of information within an organization. The use of ICT can help to dissipate information quickly and effectively. For example, if warehousing wanted to check current stock levels and pass this information on to the ordering department they could track current levels on a computer system and run off a report, instead of physically checking the stock levels.

The flow of information within an organization is very dependent on the source and recipient of the information. The source is important in terms of accuracy and translation, and the recipient in terms of their reaction to the information and the time it takes to process and respond to the information.

The recipient, in conjunction with the sender, can be a human or electronic resource. Once information has been received the recipient can respond in a number of different ways. First, they or the system can pass on this information, making them or the system both a recipient and a sender. Secondly, they or the system can carry out an action as a result of the information received, therefore activating a process. Finally, on receipt of the information they or the system can do nothing, retaining the information and storing it within their own memory bank. These actions trigger a range of information flows (Figure 3.4).

Information is sent to the receiver, who then passes it on to a third party. The recipient thus has a dual role of recipient and sender.

Information is sent to the receiver, who then carries out an action and processes the information received.

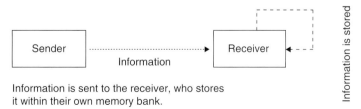

Information is sent to the receiver, who stores it within their own memory bank.

Figure 3.4 Information transmission and storage

Information flows within an organization can be tracked easily through the use of specified tools and techniques, as shown in Figure 3.5.

Figure 3.5 Information flow diagram components

Information flow diagrams provide a simplistic overview of how information is routed between different parts of an organization and also how information is communicated between the internal and external components of an organization, e.g. a student applying for a computing course at college (Figure 3.6).

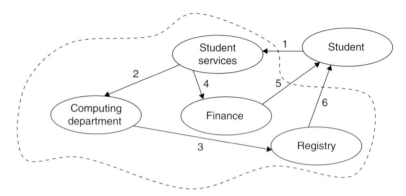

Figure 3.6 Information flow diagram. 1: Application form; 2: student details; 3: letter of course offer; 4: course fee information; 5: course fee request; 6: course induction details

Activity 3.3

1. What factors can influence the flow of information within an organization?
2. Do you think that information flows better in smaller or larger organizations?
3. Do you think that ICT helps or hinders information flow within an organization? Fully justify your answer.
4. What can happen to information when it is passed between two or more parties?
5. Give four examples of an information flow within an organization.

Activity 3.4

1. Complete five possible information transfers on the following information flow diagram.

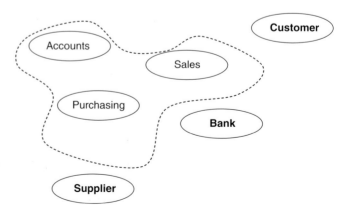

2. Identify three possible information flows that could be exchanged between the following:
   - doctor and patient
   - student and lecturer
   - parent and child
   - employer and employee.

Understand the issues and constraints in relation to the use of information in organizations

Information is vital to any organization. It provides the framework for decision making and strategy planning. There are many issues and constraints regarding the capture, processing and output of information, some of which are due to legal obligations, while others focus on ethical reasons or operational issues that can affect how an organization functions on a day-to-day basis.

Legal issues

To protect organizations, users of information systems and the general public, about whom information may be stored, a number of measures and a range of legislation can be enforced and used as a controlling measure.

The enforcement of legislation can impact on the procedures of organizations in a number of ways. In order to comply, an organization will have to ensure that they operate within certain legislative boundaries that include informing employees and third parties about how they intend to safeguard systems and any information collected, processed, copied, stored and output on these systems.

The types of legislation that an organization need to consider affect everyday operations in terms of:

- collecting, processing and storing data
- using software
- protecting their employees and ensuring that working conditions are of an acceptable standard.

Data Protection Act 1988

The Data Protection Act applies to the processing of data and information by a computer source. The Act places obligations on people who collect, process and store personal records and data about consumers or customers. The Act is based on a set of principles which binds a user or an organization into following a set of procedures offering assurances that data is kept secure.

The main principles are:

- Personal data should be processed fairly and lawfully.
- Personal data should be held only for one or more specified and lawful purposes.

CHAPTER 3

- Personal data held should not be disclosed in any way incompatible to the specified and lawful purpose.
- Personal data held should be adequate and relevant, not excessive to the purpose or purposes.
- Personal data kept should be accurate and up to date.
- Personal data should not be retained for any longer than necessary.
- Individuals should be informed about personal data stored and should be entitled to have access to it and if appropriate have such data corrected or erased.
- Security measures should ensure that no unauthorized access to, alteration or disclosure or destruction of personal data is permitted and protection against accidental loss or destruction of personal data is given.

Freedom of Information Act 2000

The Freedom of Information Act gives anybody the right to ask any public body about the information they have on any subject. An organization in the public sector will be constantly under scrutiny. For a private organization, any contracts in the public domain and any competitor information becomes more open.

Computer Misuse Act 1990

The increased threat of hackers trying to gain unauthorized access to a computer system led to the enforcement of the Computer Misuse Act. Prior to this Act there was minimal protection for individual users and organizations and prosecution was difficult because theft of data by hacking was not considered as depriving the owner.

Ethical issues

Many ethical issues may be raised with regard to the use of information in organizations. Some organizations impose a code of practice or conduct that sets outs the conditions of use for areas such as e-mail and the Internet. There may also be open policies about 'whistleblowing', in that employees are encouraged to inform of any misdemeanours concerning the misuse or abuse of IT, data security or privacy issues.

Some organizations have more general organizational policies that may cover areas such as:

- use of standard document templates
- standard document content and use of language
- confidentiality of data and information
- security of data
- communication of data and the use of passwords or encryption tools.

Information ownership is important to an organization: if an employee working for the organization has generated a document, to whom does that document belong? If information is being stored by an organization about a customer, who has the right to access it and can it be passed on to a third party? These are questions that should be addressed.

Operational issues

Various operational issues can constrain the use of information in organizations. In terms of security there are additional costs associated with setting up firewalls, installing antivirus and spyware software. A company may have moderators and additional staff to 'police' the system, and physical security can also be a drain on resources. Backups can also drain resources in that different mechanisms may have to be used, and offsite backup storage may also be required for additional security. Health and safety and compliance with legislation may require an organization to make changes to the working environment that could impinge on data requirements and storage facilities, as well as employee welfare.

Other operational issues include the introduction of organizational policies and business continuance plans that look at areas such as 'disaster recovery'.

The implication of all of these operational issues is that additional costs and resources may be required to support the introduction of new measures and procedures. Costs may include development, hardware, software and training. In addition, if more sophisticated systems are required this could impact on the skill levels of employees and require more complex hardware and software.

Know the functions and features of information systems

Information systems are systems that have been set up to manage and support the day-to-day activities of an organization and its management. Almost every organization will have information systems, ranging from a basic system relying on simple application software to process, store and deliver information, to complex, integrated systems that support the entire organization. Examples are:

- stock control and inventory
- payroll
- invoicing
- customer accounts
- ordering and distribution.

Information systems can be classified in terms of their function and complexity. General information systems use application software tools to process, store and deliver data and information. More specific information systems are used to support a very specialist function or need within an organization.

Information system tools

There are many ways in which data can be extracted to provide business information. Some of the tools that can be used to do this include the use of databases, artificial intelligence, expert systems, the Internet and

CHAPTER 3

a range of other specialist tools and systems such as data mining or predictive modelling techniques.

Some tools are quite straightforward and involve the sorting or filtering of information using conventional application software such as a database; however, there are specific tools and techniques available to serve this purpose. These knowledge tools include:

- expert systems
- data mining.

Expert systems

Expert systems represent an advanced level of knowledge and decision support systems. Expert systems encapsulate the experience and specialized knowledge of experts in order to relay this information to a non-expert, so that they too can have access to the specialist knowledge.

Expert systems are based on a reasoning process that resembles human thought processes. The thought process is dependent on rules and reasoning, which has been extracted by experts in the field. The primary function of an expert system is to provide a knowledge base which can be accessed to provide information such as a diagnosis for a patient, to assist non-experts in their own decision-making process.

Data mining

Data mining is a generic term that covers a range of technologies. The process of 'mining' data refers to the extraction of information through tests, analysis, rules and heuristics. Information will be sorted and processed from a data set in the hope of finding new information or data anomalies that may have remained hitherto undiscovered.

Data mining embraces a wide range of technologies, including rule induction, neural networks and data visualization, all working to provide an analyst with a more informative and better understanding of the data.

Information system examples

Information systems can be defined in terms of the level at which they offer support, such as strategic, tactical or operational level systems. Information systems can be used by a number of different users within an organization and a range of different functional departments, such as:

- **marketing** – for sales performance and competitor analysis
- **finance** – for financial accounting, investments and return figures
- **human resources** – for staffing data or professional development requirements.

Within each of these functional departments an information system can operate at a strategic, tactical or operational level. The levels at which an information system can function are shown in Figure 3.7.

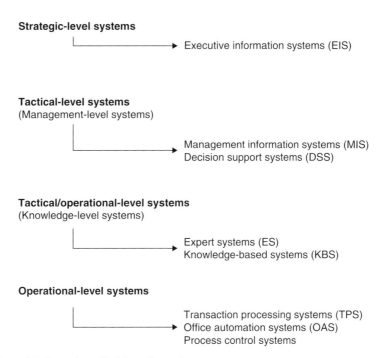

Figure 3.7 Types of specific information system

Strategic-level systems

This level of information system supports senior executives in making unstructured decisions at a strategic level. The types of decisions that could be made include:

- Should we consider diversifying into new markets?
- Should we make a bid to acquire new businesses?
- How could we embrace new challenges in the area of e-commerce?

Strategic level information systems are set up to forecast, budget and plan for the future, extending over the long term, a period of five years and beyond. Within this category specific information systems can be set up, for example executive support systems (ESS).

Executive support systems

ESS exist to support strategic personnel within an organization, their function being to provide the support and guidance needed to carry out long-term forecasting and planning. ESS use data and information collected from the current environment to establish trends or anomalies which can then be used for future planning. For example, an organization that may wish to transfer production to Europe over the next five years may look at a range of available data sources, including:

- cost of manufacturing (labour, transportation, premises)
- import and export issues (cost, initiatives, barriers to trade)
- existing businesses already trading in Europe and their profitability
- current financial status and whether there would be enough capital to finance such a venture in the future
- existing competition in Europe.

In order to identify specific trends, ESS may also rely on historical data to identify what has been done in the past and whether it was successful.

A successful ESS will have the characteristics shown in Figure 3.8.

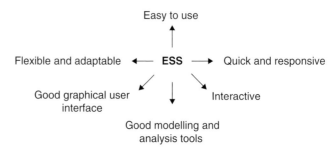

Figure 3.8 ESS characteristics

Users who will be accessing ESS may have very limited IT knowledge or skills, senior executives will not necessarily be technically orientated and therefore the ability to access the ESS easily and quickly is essential. Information required should be provided by the ESS within a specified period to enable further decisions to be made quickly.

An ESS must be able to interact easily and effectively with other systems in order to retrieve the data required. For example, decisions to be taken on whether or not to take over a new company may require the ESS to retrieve financial share price data from an external database source such as the London Stock Exchange. So that the correct decisions can be made, the modelling and analytical tools should be first class and the graphical user interface (GUI) must to be easy to use, visual and instructive.

Finally, an ESS has to be flexible and adaptable in order to continually support the ever-changing requirements of an organization.

Tactical/management-level systems

Tactical/management-level systems are designed to support middle management in their role of making some unstructured and semi-structured decisions, but of a lower level than those offered by strategic-level systems. These systems are put in place to offer support to management levels within an organization; they are not exclusive to managers. Tactical/management-level systems provide a category in which other information systems are embedded, including:

- management information systems
- decision support systems
- operational information systems.

Management information systems

Management information systems (MIS) support management at all levels within an organization by providing them with data and information based on both current and historical records, from which informed and detailed decisions can be made. MIS is typically based on internal data. Examples of this include:

- financial status
- performance and productivity levels
- weekly, monthly, quarterly forecasts and trend analysis
- sales targets and figures.

The primary role of an MIS is to convert data from internal and external sources into information so that it can be communicated to all levels within an organization. Management levels will use the information produced to enable them to make more effective decisions.

Decision support systems

Decision support systems also support management levels within an organization, helping them to make dynamic decisions that are characterized as being semi-structured or unstructured.

These systems must be inherently dynamic to support the demand for up-to-date information, enabling a fast response to the changing conditions of an organization. Decision support systems are complex analytical systems that are designed explicitly with a variety of analysis and modelling tools to process, enquire about and evaluate certain conditions.

Tactical/operational-level systems

Tactical/operational-level systems are specialist systems that provide support for knowledge users within an organization. This particular category of information system is not confined to a specific user, e.g. manager, or a specific decision type, structured, semi-structured or unstructured.

The function of this type of information system is to assist an organization in their quest to:

- identify
- discover
- analyse
- integrate
- collaborate

new ideas and information, to make the organization more efficient or profitable, or to ensure high-quality standards among the workforce, services offered and/or production lines.

The users of this level of system are generally those who have achieved high academic degree or further degree status, or are members of recognized professions such as engineers, doctors, lawyers or scientists; their role within the organization being to seek out technical facts, information and knowledge, which can then be analysed, processed and integrated into the organization. Examples of how these systems can be used in a hospital include:

- identification of certain patients who are more at risk of certain medical conditions

CHAPTER 3

- the impact of certain drugs on certain categories of patients
- the impact of monitoring close relatives' medical history on patients.

Operational-level systems

As illustrated in Figure 3.7, this level of information system supports operational managers and supervisors, and assists them by tracking and monitoring activities that occur at this level. The categories of system that come under the operational level include:

- transaction processing systems (TPS)
- office automation systems (OAS)
- process control systems.

The types of activities that may occur at this level include:

- sales figures for a set period
- production and productivity levels
- ratios examining daily work flow.

A system at this level will answer routine questions such as:

- How much is being produced on a certain basis?
- How many items are in stock?
- When will production targets be met based on current workflow levels?

Operational-level systems will provide answers to structured questions and decisions where there may be a limited number of outcomes. For example:

- How many items are in stock?

Stock Report as at 1 March 2002			
Stock Number	Stock item	Quantity	Location
RT1244000	Fan belts	136	Aisle 6B
Y45501	Spark plugs	26	Aisle 2A
FG2670911	Fuses	12	Aisle 1D
HI611098	Washers	180	Aisle 1B

There are three different types of information in the operational-level category:

- transaction processing systems (TPS) or data processing systems (DPS)
- office automation systems (OAS)
- process control systems.

Transaction processing systems/data processing systems

These systems exist to support the operational level of organizations and assist in providing answers to structured routine decisions. TPS is pivotal to any organization because it provides the backbone to day-to-day activities and processing. Examples of TPS include holiday booking systems, customer ordering systems and payroll systems.

Data processing systems

Data processing systems carry out the essential role of gathering, collating and processing the daily transactions of an organization. These systems are also referred to as transaction processing systems (TPS). Typical functions of a TPS include:

- accounts
- invoicing
- stock management
- ledger keeping.

TPS is an essential part of an organization because it keeps its operations and day-to-day activities running smoothly and provides a base for other information support, including MIS.

Data processing systems can be characterized by their prespecified functions in that their decision rules and output formats cannot easily be changed by the end-user. These systems are directly related to the structure of an organization's data.

Office automation systems

These systems are set up to identify and increase levels of efficiency and productivity among the workforce. To assist in this role, various tools and software are available to schedule, monitor and improve workforce activities. OAS will enable the workforce to:

- communicate more effectively
- promote collaborations and group synergy
- structure daily tasks and activities
- track and schedule appointments and activities
- increase productivity by reducing repetitive workload
- automate repetitive tasks.

OAS can be quite simple, drawing on the functions of application software such as word processors, spreadsheets, databases, multimedia and communications software such as e-mail.

More complex and software tools can also be used to focus on a specific area of workflow or productivity, such as GroupWare, document imaging processing, workflow management systems or electronic document management systems.

Process control systems

Process control systems monitor, support and control certain process activities within a manufacturing environment. Applications that are used to support process control systems can help an organization in the following ways:

- improving quality control
- assisting with project planning of the product
- assisting with physical design
- identifying resource requirements
- identifying development status or stage in the product life cycle.

CHAPTER 3

A wide range of software is available to support both general and specific activities that fall under the domain of process control systems, as identified in Table 3.4.

Table 3.4 Support available for process control systems

Software type	Function
Spreadsheets	Costing of manufacturing items
	Forecasting sales
	Identify break-even and profit margin points
	Analyse work patterns and efficiency levels
Statistical packages	Examine productivity levels to identify ratios of optimum working conditions
	Identify relationships between workforce and productivity
Project management	Uses Gantt charts to identify timings of activities
	Schedules tasks and activities
	Identifying task dependencies
Computer-aided design (CAD)	Interactive development of drawings and designs
	Professional drafting tool
Computer-aided manufacture (CAM)	Controls production equipment more accurately
	Integrates with other manufacturing systems
	Ensures quality procedures

Management information systems

MIS can be quite complex in that they have a number of features and benefits, and are based on a range of effectiveness criteria, such as their accuracy, sustainability, response time and the confidence that they generate among users.

MIS summarize and report on the basic operations of an organization. The MIS converts data from a variety of internal and external sources, usually via a TPS, and presents the output information in an appropriate format such as a report that can be used by managers at different levels (Figure 3.9).

MIS are used to support tactical and strategic decision making, and are therefore associated with use by managers at these levels. MIS embraces a range of information systems, such as:

- information reporting systems
- decision support systems
- executive information systems.

To assist managers in their decision making, data and information need to be made available in an appropriate format. Information reporting systems provide reports to meet this purpose. Reports can be generated periodically at predetermined intervals, e.g. every hour, day, week or month, on demand, as and when required or when an event triggers the need for a report.

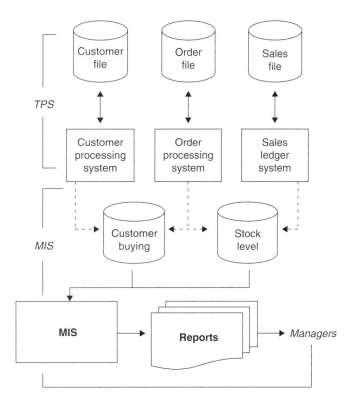

Figure 3.9 MIS input, conversion and output system

The report will be the result of data collected internally and possibly externally at an operational level, on the shopfloor, processed and output information reporting systems. Typical report types are shown in Table 3.5.

Table 3.5 Report types

Report type	Internal data	External data
Stock requisition	Current stock levels and stock prices (generated by warehouse or stock personnel)	Availability and pricing of stock items (suppliers)
Productivity	Number of widgets made in a set period, number of personnel making the widgets and cost of production	Competitor widget output levels, market share information

Decision systems or decision support systems (DSS) provide managers with the knowledge and information needed to support semi-structured or unstructured decisions.

The way in which DSS work is by mimicking the way in which actual human experts would go through the process of decision making. Using tools and software such as artificial intelligence, fuzzy logic, data mining, knowledge-based systems, neural networks and heuristics, the DSS will learn and store up a knowledge base to support the decision-making process.

The function of executive information systems (EIS) is to provide analytical, comparative and forecast tools to enable effective strategic

CHAPTER 3

decision making. EIS are designed to support senior managers; however, information may be generated at an operational level to make informed and accurate decisions.

Activity 3.5

Most organizations have MIS support. Your task for this activity is to carry out research and find an organization that uses an MIS system. Once you have found one, answer the following questions.

1. What is the role of the MIS within the organization?
2. What specific areas or functions does the MIS target and support?
3. What system was in place before the MIS (if any)?
4. Prepare a short report on the MIS system that can be presented orally to the group.

Elements of information systems

A number of key elements can influence information systems within organizations. These elements can be categorized into the following main areas:

- data
- human resources – people
- hardware and software resources
- telecommunications.

Data

An information system is really only as good as the data that is fed into it for processing purposes. If poor data is captured, then the processing will be poor and the output results will also be of a poor quality. If data sets are incomplete or inaccurate then the final information will also be inaccurate and incomplete, rendering the information system quite useless.

Human resources – people

The people within an organization heavily influence information systems. Human resources are required to manage and maintain the data that supports the information system and the decision-making process that is pivotal to strategic-based systems.

The design of information systems is also dependent on the knowledge and skills of the personnel who manage, programme, analyse and implement the technology. Depending on who has had input into the overall design, issues of ownership may impact on and influence the functionality of the information system by restricting its use to support individual levels, e.g. management, within the organization.

Other factors can influence information systems development, including the way in which raw data is collected, the type of processing activity,

Open system – a system that interacts with other systems or outside its environment.

Closed system – a system with little or no interaction with other systems or the outside environment.

cost and time issues and dependence on human operators and designers. To ensure that information systems are used at optimum levels a balance of all of these resources needs to be considered.

Hardware, software and telecommunications

The information system is, to an extent, constrained by the technology used to support it. If the hardware and software combination is modern, with a high specification and good storage facilities, and is fast and user friendly, then it can make the overall processing and output of information more dynamic.

The type, speed and mode of telecommunications used can also influence the information system in terms of quality, performance and accuracy.

Information system functions

Information systems perform a number of functions, including the input, storage, processing and output of information. They also provide control and feedback mechanisms (Figure 3.10) to support the end-users in either an open or a closed system environment.

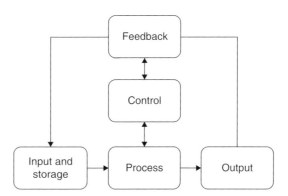

Figure 3.10 Information system functions

Information can be input through a number of devices, such as handheld scanners, readers, keyboards, data transfer tools and download mechanisms. An information system can then store the data within a piece of software, which can then process the information and output it in a suitable format such as a report. Feedback and control mechanisms can be built into the information system to alert the user to any errors in the data, structure or processing activity. All of these functions can operate within an open or a closed environment.

References

O'Brien, J. (1993) *Management Information Systems: A Managerial End User Perspective*, 2nd edition, Richard D Irwin, Boston, USA.

CHAPTER 3

Questions and review

1. Characterise the differences – with examples – between data and information.
2. Information sources can be defined as being internal or external. Provide examples of both sources.
3. Organisations use business information for a range of purposes. In terms of operational support, give two examples of how this can be used to support an organisation at this level.
4. How can information help an organisation gain competitive advantage? Carry out research to identify an organisation that has done this.
5. Provide detailed descriptions of the main functional areas of an organisation.
6. Produce an information flow diagram for a system or organisation that you are familiar with.
7. There are a number of issues and constraints relating to the use of information in organisations. One of these issues relates to the legality of how information is used and stored. A range of legislation has been set up to protect, monitor and enforce penalties if breaches are made.

   Can you identify two pieces of appropriate legislation and describe how they are used to ensure organisational compliance?
8. What ethical issues need to be considered in terms of the use of information in and by organisations?
9. There are a range of operational issues in relation to the use of information in organisations. Keeping information secure is of vital importance to every organisation. Can you identify ways in which you can keep information secure on a networked system?
10. There are a range of information system tools such as: artificial intelligence, expert systems, the internet, data mining systems and predictive modelling tools. List the features of three of these tools in relation to information systems.
11. Can you provide two examples of information systems and how they are used?
12. What are the features and benefits of an MIS?
13. There are five key elements to information systems, what are they?
14. What are the primary functions of information systems?

Assessment activities

Grading criteria	Content	Suggested activity
Pass		
P1	Describe the characteristics and sources of information that an organisation needs.	Produce a report that describes the characteristics and sources of information that an organisation needs.
P2	Describe how information is used for a range of purposes in a selected organisation.	Include an additional section within the report identifying how information is used for a range of purposes in a selected organisation.
P3	Describe how information flows between different functional areas in an organisation.	Using annotated flip chart paper, describe how information flows between different functional areas in an organisation.
P4	Describe the features and key elements of a management information system showing where it supports the functional areas of an organisation.	Produce further sheets that illustrate and describe the key elements of a management information system, showing where it supports the functional areas of an organisation.
P5	Identify the constraints that relate to the use of customer information in an organisation and describe how these may impact upon the organisation.	Produce an information leaflet that could be used for new employers as part of their induction to an organisation. The leaflet should identify the constraints that relate to the use of customer information in an organisation and it should also describe how this could impact upon the organisation.
P6	Describe different tools used to manage and process information.	Produce a presentation that describes different tools used to manage and process information.
Merit		
M1	Explain the importance to an organisation of effectively collecting, processing and using information.	In conjunction with P1 and P2, include an additional section within the report that explains the importance to an organisation of effectively collecting, processing and using information.
M2	Compare, using examples, the usefulness of different tools for processing information to support effective business decision-making.	In conjunction with P6, prepare a presentation that compares – using at least three examples – usefulness of different tools for processing information to support effective business decision-making.
M3	Explain the purpose and operation of data mining and predictive modelling.	Design a brochure to be presented at a data and information processing conference. Within the brochure give explanations as to the purpose and operation of data mining and predictive modelling.
Distinction		
D1	Explain how an organisation could improve the quality of its business information justifying each of their recommendations.	In conjunction with P1 and P2, add a further section to the report that explains how an organisation could improve the quality of its business information, justifying each of the recommendations put forward.
D2	Evaluate tools used for processing information with respect to their support in decision making.	In conjunction with M2, include additional slides that evaluate tools used for processing information with respect to their support in decision making.

CHAPTER 3

Courtesy of iStockphoto, Branislav, Image# 4564045

IT systems analysis and design provides a comprehensive and quite prescriptive approach to addressing a range of business projects and solutions. Based upon a given methodology a certain life cycle can be followed that will provide a holistic approach to analysing, designing and implementing systems.

Identifying a need for change within an organisation that could be driven by technology, resources, growth, costs, performance or efficiency; and being able to provide a solution that addresses this need, is one of the merits of systems analysis. The ability to resolve and improve, provides the foundation to this process, and if implemented correctly and accurately can support a business in becoming more competitive, cost-effective, profitable, compliant or productive in terms of meeting its objectives.

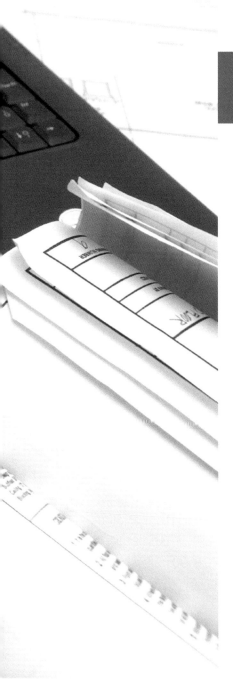

Chapter 4

IT Systems Analysis and Design

S ystems analysis provides a very structured methodology that can help analysts and designers to build systems that reflect the needs and requirements of end-users. The principles applied in systems analysis are based on a life cycle that outlines how data and information can be captured, processed and stored. It also identifies various models that can be used to analyse information and design techniques for building an appropriate systems solution. The systems analysis life cycle also examines strategies for implementation, testing, and review and maintenance.

This chapter will provide you with an overview of the tools and techniques used in systems analysis, and with the knowledge to apply these to a range of case study and activity-based scenarios.

Each section within the chapter will focus on one of four learning outcomes:

- Understand the principles of systems analysis and design.
- Be able to investigate, analyse and document requirements.
- Be able to create a system design.
- Be able to design a test plan.

Embedded within each section is a range of activities that will strengthen your understanding of the subject matter and provide you with the support you need to complete effectively some of the evidence requirements for the unit.

Understand the principles of systems analysis and design

The systems analysis life cycle has evolved from the structure set out in structured systems analysis and design methodology (SSADM). Systems analysis and design (SAD) fits into this structure; however, it is less prescriptive and the stages involved in the life cycle are more informal in terms of identification of specific steps and tasks. SAD has evolved over the years into a methodology that is quite dynamic and effective in terms of determining, analysing and providing feasible solutions to a range of systems environment problems. SAD has been developed around a user-friendly and adaptable life-cycle framework that can be applied at any level to almost all types of organization and end-user environment.

Development life cycles

Along with the SSADM or SAD life cycle, other development models exist, including:

- the waterfall model
- rapid applications design (RAD)
- the spiral model
- dynamic systems development methodology (DSDM).

The waterfall model is based on a series of steps that should be addressed when building an information system. The order of these steps is predefined, with a review at the need of each. The sequence of phases in the waterfall model is shown in Figure 4.1.

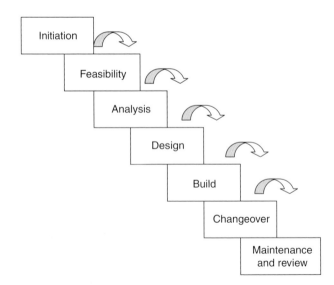

Figure 4.1 Waterfall model

RAD uses prototyping to develop information systems, with the added benefit of achieving faster development than the more traditional methodologies.

CHAPTER 4

The spiral model is an iterative systems development model that is based on four main activities:

- planning
- risk analysis
- engineering
- customer evaluation.

This model was developed in response to the fact that systems development projects tend to repeat the stages of analysis, design and code as part of the prototyping process.

DSDM is based on the principles of RAD. Put together by the DSDM Consortium (http://www.dsdm.org), DSDM is a more up-to-date version of RAD. The result of this approach is that new systems are built more in line with the needs of the intended users.

The benefits of a life-cycle model are that they provide a strong framework on which to plan, design and implement a system or project. The life-cycle models are quite robust; they follow set procedures that are to an extent incremental, thus ensuring that the system is designed in the correct way using a correct, almost prescriptive, process. Each stage within a life-cycle model is of significance because it involves a set of tasks that offer support both to previous tasks and to the tasks that will follow. An example of this can be seen with the systems analysis life cycle, as shown in Figure 4.2.

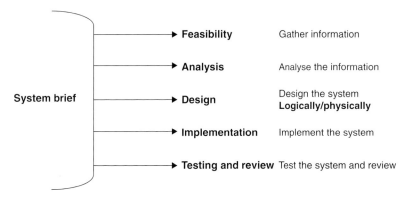

Figure 4.2 Systems analysis life cycle

Development methodologies

Development methodologies or ways of doing things, structuring, designing and building systems are required to ensure that the system being built has followed a structured walk-through process.

The tools and techniques that can be used include:

- activity diagrams
- data flow diagrams
- computer-aided software engineering (CASE) tools

All of these use certain tools and procedures to facilitate the understanding, analysis and design of a system.

Activity diagrams

An activity diagram illustrates the step-by-step workflow components/elements within a system (Table 4.1). The components could be of a business or an operational nature.

Table 4.1 Components of an activity diagram

Initial activity – a solid circle represents the start of the first activity	
Activity – this is represented by a rounded rectangle	Description of the activity
Decisions – a diamond represents the point at which a decision is to be made. The options from that decision are represented as arrows that emerge from each side of the diamond with a written description identifying that option	
Signal – polygons are used to represent signals (input and output). When a message is sent or received by an activity this is called a 'signal'	Input Output
Concurrent activities – these run simultaneously or occur in parallel with others. An example is talking to a friend on the phone about a holiday and checking out a website on flights. To represent this a bold horizontal line is used to mark the beginning and end of the concurrent activities	Talk View
Final activity – a target board symbol denotes the final activity in the diagram	

Data flow diagrams

Data flow modelling is an established tool that is used to examine the environment of the system under investigation through the use of data flow diagrams (DFDs) and associated descriptors which are used to identify and establish the following:

- the flow of information within a system
- the processing activities that take place
- the storage mechanisms used.

Together the diagrams and the descriptors provide a top–down approach to understanding the system (Figure 4.3).

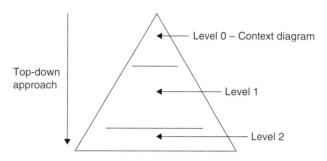

Figure 4.3 Top–down approach of data flow diagrams

The top–down approach adopted in data flow modelling is designed to provide a more in-depth analysis of the system the further down the levels. At level 0, which is also referred to as a context diagram, a single process is used to identify the flows of information between the system in its entirety and any external entities.

Level 1 DFDs use a number of processes to represent what is happening in the entire system under investigation. The level 1 diagram provides an insight into the processing of information, sources and recipients of information, types of information and the storage mechanisms used, the diagram providing a visual insight into the system. An example of a level 1 DFD can be seen within the TEY Supermarket case study on page 107.

Level 2 diagrams are specific to certain processes that have already been identified at level 1. Level 2 diagrams use a single process which is a detailed expansion of the version represented at level 1. Within the single process more specific information is detailed which provides a true and more accurate representation of the system. An example of a level 2 DFD can be seen within the TEY Supermarket case study on page 107.

CASE tools

CASE tools are considered to be of great importance on large-scale projects. The main features of CASE tools are that they can control and help to document large projects. These tools are designed to support diagram creation and any amendments through a series of consistency checks. Some uses of CASE tools are:

- as tools for diagram creation
- as tools for validating diagrams
- for generating low-level diagrams
- for generating reports
- for generating codes.

Like any software tool, CASE tools can supply the diagrammatic, editing, processing and aesthetic framework for the project; however, the project content will be down to the user.

Key drivers: the need for systems analysis

Systems analysis is used to support projects and systems development for several reasons, some of which are based on a business need

(Figure 4.4). All businesses have needs that may be based around their company aims and objectives. Some of these may be based on the need to grow and expand: diversification into other markets and physical growth. Expansion and growth through company acquisitions is also a way of achieving a greater market share and possibly becoming more competitive, especially if the company acquisition involves buying smaller or competitive rivals.

Figure 4.4 Need for systems analysis

The need to be more efficient and productive is another need that is especially important to service industries that are heavily focused on customer service. Increased productivity and efficiency could result in higher profits, more motivated staff or enhanced competitiveness in the marketplace.

Sometimes a legal requirement may generate a specific business need. For example, compliance with health and safety legislation may generate the need for better hardware and software provisions, more ergonomically designed offices or rescheduling of workloads.

Benefits of effective system analysis procedures

Systems analysis can be very instrumental in the success of an organization and its business procedures if implemented correctly. The effective use of systems analysis procedures can result in risk reduction concerning projects running over budget and costs spiralling. In addition, the final system and associated hardware or software should meet the requirements of the user. Finally, any system that is designed should be easily maintained, upgradable, robust and dynamic.

Be able to investigate, analyse and document requirements

Systems analysis follows a life cycle that allows for investigation, analysis, design and documentation to take place seamlessly. By following the procedures at each stage, information required to make value judgements about the revised or new system will be valid, fully justified and appropriate.

CHAPTER 4

Investigative techniques

Various investigative techniques can be used to capture information in the feasibility/fact-finding stage of a systems analysis investigation. These include:

- interviewing
- questionnaires
- meetings
- observation
- documentation analysis.

In addition, there are several factors to take into consideration when conducting a fact-finding investigation. These factors can affect how data is analysed, and issues of sensitivity when collecting information and when observing users in their work environment. The use of a cost–benefit analysis can also be applied as an investigative technique to analyse the requirements of a proposed system and assess its feasibility.

Interviews

Interviewing users of a system is one of the main investigative techniques that can be used in a fact-finding investigation. Some of the objectives of interviewing are shown in Figure 4.5. Interviewing may uncover new information and provide an opportunity to understand the system better through the eyes of the user.

Figure 4.5 Interview objectives

A number of factors should be considered when interviewing users (Figure 4.6):

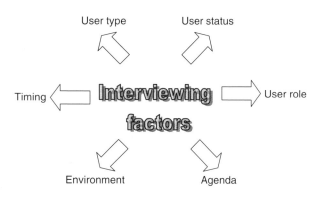

Figure 4.6 Interviewing factors

- **User type** – are they senior management, head of a department, team leader, data-entry clerk or administration staff, etc.?
- **User status** – people higher up in the hierarchy of the organization may have more limited time to assist with an interview.
- **User role** – what position do they have within the organization and what impact do they have on the system investigation?
- **Agenda** – users within the system may have their own reservations about changes to their system and may therefore be biased in their interview answers. It is the role of the analyst to decide what is fact and what is fiction. This can be achieved by validating information with a third party.
- **Environment** – will influence the quality of information given by a user. Users will feel more comfortable in certain environments in which they feel safe and are sure that they can talk in privacy.
- **Timing** – crucial in terms of the qualitative aspect of the interview. If the interviewee is prepared, and has made an effort to set time aside for questioning without interruptions, the information given will be of a better quality and more detailed.

Activity 4.1

In pairs set up a fact-finding interview to identify what your interview partner did at the weekend. One person will provide the information about what they did and the second will ask a serious of questions to extract as much information as possible.

To ensure that both people have an interviewing role, once the first interview has been conducted, the roles should be reversed so that the interviewee becomes the interviewer.

At the end of the exercise information should be fed back to each interviewee. During the feedback stage the interviewee should say how they felt during the interview process and what information was not uncovered. The interviewer should then feed back and confirm that the information extracted was correct.

Questionnaires

Questionnaires are an excellent way of gathering and consolidating information, providing the following conditions are met:

- The questionnaire is structured appropriately.
- A control mechanism is in place for gathering the questionnaires.
- The correct user group has been targeted.

The questionnaire should be set out clearly to provide opportunities for short answers based on facts and figures, giving the quantitative aspect, and descriptive answers providing the qualitative aspect, as shown in Figure 4.7. A balance of questions will ensure that all of the information required is gathered to continue with the investigation.

ID number:	001	System Objective: **Upgrade computers in the Finance and IT departments**	
Name: Robert Smith		Department: Networking	Job Title: IT Support Administrator

Tasks undertaken each day:

- Remove backup disks and take them offsite
- Set up new users on the system
- Set up security on the file systems
- Produce system documentation and procedures manuals
- First line support – help desk
- Assist with installations and upgrades

Communicates with: Network manager, other IT support staff in the department, users at all levels, software and hardware manufacturers

Documents used:

- New user setup forms
- Internet access forms
- Backup schedules
- Support call log

Constraints and problems:	User solutions:
1. Too much documentation	Make the support call log automated using a database
2. Users sometimes have to make multiple requests for passwords because there is no tracking system of who applied when, and sometimes the setup forms get mislaid	Better storage system and introduce a tracking system

Please tick the following if you agree:

Problems exist with the following

Network ☐   Operating system ☐   Other software ☐   Inexperienced users ☐

Please identify how the above have contributed to the problems with the IT system:

Any other information:

User complaints about the time it takes to attend a callout. Network keeps crashing especially between 8:00 and 9:00 in the morning

Figure 4.7 Sample questionnaire template

It is always best to provide a time limit for the return of questionnaires; for example, 'Please return within three working days'. Another way to ensure that the questionnaire is returned is to ask users to fill them in, then collect them at the end of a focus group session or meeting.

When designing a questionnaire another factor to consider is who the questionnaire is aimed at. The target audience is very important because different users can interpret a question very differently depending on their status and the role they play within the system.

CHAPTER 4

Activity 4.2

1. Design a questionnaire that could be given to one of the following users:
   - the manager of a new company that sells records and CDs online over the Internet
   - a checkout operative at a supermarket
   - a teacher/lecturer working in a school or college.
2. The aim of the questionnaire is to identify what the user does within their system, how they do it, who they communicate with and what pressures or constraints exist within their system.

Activity 4.3

1. Set up small groups based on a mix of users for which the questionnaire was designed. Each group should have at least two of the chosen user types, i.e. manager, checkout operative or teacher/lecturer.
2. Within the group identify which questions are common to the range of users and which questions are unique to a specific user. Discuss why these similarities and differences occur.

Activity 4.4

1. What are the four main fact-finding techniques?
2. Identify two other techniques that could also be used.
3. What factors should be considered when interviewing users of a system?

Meetings

Sometimes it is more cost-effective and less time-consuming to speak to users en masse by setting up a focus group, seminar session, question and answer group or general departmental meeting.

Observation

Depending on the type of system under investigation, observation may be the primary tool for gathering information. Observation is especially effective in dynamic environments where lots of activities are taking place. These activities may not rely on masses of documentation and because of the nature of the activity interviewing or questionnaires may not be appropriate. In this scenario an analyst could observe the users and then record what is being achieved and how.

An example of this is the observation of a waiter or waitress in a restaurant who is under pressure to serve the customer quickly and efficiently. Time may be limited in which to conduct an interview, the balance of qualitative questions on a questionnaire may not be sufficient to draw conclusions from and the only document in use may be a food order pad.

Documentation analysis

Documentation analysis is a very good way of understanding the way in which a system operates and the processing activities that take place.

Through investigation of documents an analyst can authenticate user statements as to what happens within the system and how it happens, and trace the source and recipient of certain information.

Examining physical data and information will also ensure that the facts are correct because they have been documented formally. Documents that could be examined in an investigation include:

- invoices
- purchase orders
- goods received notes
- receipts
- customer records.

Activity 4.5

For each of the following systems identify five documents that could be examined and describe how important this information would be to an analyst:

- travel agents
- doctor's surgery
- school or college
- supermarket
- football club.

Examination of documents will reveal how information flows internally and externally through the system, the quality and frequency of the information, and the data capture and storage mechanisms.

Activity 4.6

1. What are the benefits of investigating documents in a system?
2. What problems may arise during an investigation of documentation?
3. What document is used to present the findings of the feasibility study?

A number of factors and influences can impact on data collection and the use of investigative techniques.

Data analysis is an important part of the feasibility stage of any systems investigation. Analysing the data collected can then support any decision taken in regard to the new system proposal. Sensitivity in collecting information and observing users is crucial. If users feel threatened when they are being interviewed they may hold back on certain information that could be critical to the future success of the system. Being sensitive and empathic towards users in an interview or observation session may put their minds at rest and make them more approachable in terms of gaining access to the information required. Observing individuals at work can be quite obtrusive, especially if they are not aware of the reason why they are being observed; therefore, steps should be taken to make them fully aware of the purpose of the data collection activity.

A cost–benefit analysis is always a useful model to apply to any project situation. This model will provide a framework for identifying the costs involved and offsetting them against the associated benefits.

Requirements specification and analysis tools

Specifying the requirements of a new system marks the transition from the earlier stages of data collection, investigation and analysis, towards systems design and build.

The requirements specification stage consists of a number of tasks that examine the scope and boundaries of the proposed system, the required inputs, outputs and processing activities. In conjunction with the modelling tools and techniques used, and through the application of business system options (BSOs) and technical system options (TSOs), alternative system solutions may be generated.

These alternative system solutions, if BSO based, may look at systems that are more user orientated and driven, rather than technically based and driven. The system option may include more or additional support for end-users, such as training and possible input into the overall design. The software for the system may have a more user-friendly interface as opposed to a more technical interface.

Alternative systems based on a TSO model may include recommendations for a networked environment and the type of topology to be used. The proposed systems may include variances in cost, specification or technology. In addition, alternative systems could be proposed based on implementation strategies and the time it would take to physically install and configure the system.

Analysis tools that can be used to support the requirements specification include:

- user catalogues
- requirements catalogues
- BSOs and TSOs
- data flow diagrams (DFDs)
- entity relationship diagrams (ERDs)
- input/output structure diagrams.

Each of these tools should help you to make a more informed choice about the system solution in terms of how it will be designed, how it meets the requirements of end-users and implementation strategies.

Be able to create a system design

Creating a system design is based on informed choices that have been supported by a range of documentary evidence that has been collected throughout the fact-finding/feasibility stage of the project. Documentation is crucial to the development and success of any system as it provides the evidence to support the changes that are required to ensure that a new system is more productive, efficient or cost-effective, etc.

Documentation

A range of documentation is used within the systems analysis process. Some of this documentation is associated directly with a particular life-cycle stage, for example investigative techniques applied in the feasibility study to capture user and system information. Some documentation is related directly to the use of a particular modelling tool or techniques used, such as DFDs or ERDs. The documentation will include the actual model(s) and accompanying data dictionaries, process descriptors and other explanatory information. Other documentation is there to support a design, chart or process.

Depending on the type of system that is being designed, documentation may arise from the inputs or outputs from that system or proposed system, such as screenshots or reports, especially if a database solution has been selected.

Data flow diagrams

Four tools are used in the preparation of a DFD, as identified in Figure 4.8. The tools are used together to create a DFD. Closer examination of the tools reveals that each has an equally important role in examining different aspects of the system.

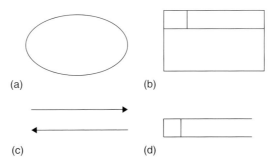

Figure 4.8 Data flow diagram tools: (a) external entities; (b) process boxes; (c) data flows; (d) data stores

- **External entities** – identify people and organizations outside the system under investigation.
- **Processes** – represent the activities that take place within the system.
- **Data flows** – provide the physical link between data sources that flow to, from and within the system.
- **Data stores** – detail the type of storage mechanism used to hold the data/information within the system.

External entities

External entities are used to represent people or organizations that have a role in the system, but are not necessarily part of the system.

Example   A hospital has called in a systems analyst to identify any areas within the hospital that are running inefficiently. The hospital is due to have inspectors in to examine the hospital's procedures and also their spending to enable new targets to be set for the next financial year. The areas of the hospital that are under investigation are X-rays and accident and emergency. The first system that you have been asked to investigate is X-rays.

Typical external entities for X-rays may include:

- patients coming into the X-ray department
- other hospital departments, such as accident and emergency or surgery
- doctors and medical staff from other departments
- inspectors.

External entities can appear in a system more than once, for example a patient would be part of a number of processes within the X-ray system, in terms of providing patient details, having a consultation with the doctor and having the actual X-ray. To indicate that a patient appears more than once in the system, this is represented by crossing the corner of the external entity, as shown in Figure 4.9.

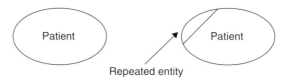

Figure 4.9 Sample external entities

Activity 4.7

1. Identify a range of external entities for each of the following systems:
   - airline company
   - mail-order catalogue company
   - wages department.
2. Identify one external entity for each of the systems which could be repeated and describe why.

Processes

Process boxes represent activities that take place within the system and activities that are linked to the system. All activities have a process attached to them; something triggers the process and an action may become the output of the process.

Example   Activity of sending and receiving e-mail

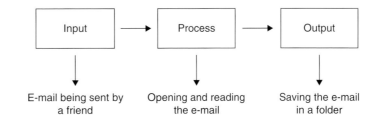

The process box has three distinct sections, each with its own identifier:

1. Identifies the process box with a unique number.
2. Stock room – provides details of the location where the activity is taking place.
3. Check items in stock – identifies the activity that is taking place.

Activity 4.8

Think about an activity that you did at the weekend and record the activity using a set of processes labelling each accordingly as shown in the example.

Example   Processes to identify on a trip to the supermarket

Data flows
Data flows indicate the direction or flow of information within the system:

Data flows provide the links to other data flow tools within a system, including those shown in Table 4.2.

Table 4.2 Data flow links

	Data store	External entity	Process
Data store	✕	✕	✓
External entity	✕	✕	✓
Process	✓	✓	✕

The matrix clearly defines that the flow of information within a system must always evolve around a process. The direct flow of information between processes is stating that two activities can take place without the intervention of an input such as a data-entry clerk. This is true in terms of automatic processing, where a programme could receive a set of information, collate or process the information, which could then automatically trigger a second process, for example running off a report. Without the use of automated systems direct links between processes would rarely exist and should therefore not be linked on a DFD.

All data flows should be labelled clearly to identify the type of data or information that is being passed to and from sources and recipients.

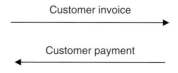

Data stores

Source – somebody or something which is the source of information (from which information flows).

Recipient – somebody or something that receives information (to which information flows).

Data stores represent different types of storage mechanisms. There are four different types of data stores:

- **D** – digitized or computerized storage mechanism, such as a file on a database
- **M** – manual storage mechanism, such as a filing cabinet
- **T(M)** – manual transient data store; a temporary manual storage mechanism such as an in-tray on a desk
- **T** – computerized data store which is temporary, e.g. e-mail which may be read once and then moved to a permanent storage file or deleted.

There are two components to a data store:

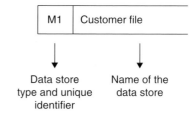

The information provided tells us that the data store type is manual, and the identifier (1) is associated with the data store mechanism, which is a customer file. If the customer file was accessed again within the system, it would then become a repeated data store. All of the information remains the same; however, we identify the repeated aspect by inserting a second line:

Repeated data store

Activity 4.9

Identify four typical manual and one typical computerized data store for each of the following systems:

- hotel reservation system
- buying a ticket for a concert
- organizing a holiday.

Designing data flow diagrams

All of the four DFD tools identified complement each other in providing an accurate, diagrammatic overview of the system under investigation. As previously mentioned, DFDs provide a top–down approach to modelling. This allows the analyst to interpret the system through three representations, the context diagram, level 1 DFD and level 2 DFD. To provide a more realistic and more importantly consistent overview of DFDs, TEY Supermarkets will be used as a case study system.

Activity 4.10

1. What four tools are used in the preparation of data flow diagrams? Describe the purpose of each.
2. Which two tools can be identified as being repeated?
3. What does the notation (D), (M) and (T) mean?

Case study 4.1

TEY Supermarkets

TEY Supermarkets is an established chain of supermarkets that are located across the country. Over the past six months the managing director of the chain, Mr Thomas North, has discovered that they are losing their proportion of the market to another competitor. Since the beginning of the year their market share has fallen from 16 per cent to 12 per cent.

TEY Supermarkets has fifteen stores across the region, all located in major towns or cities. The structure of the company is very hierarchical, as shown in Figures 4.10 and 4.11, with the lines of command being generic across all branches:

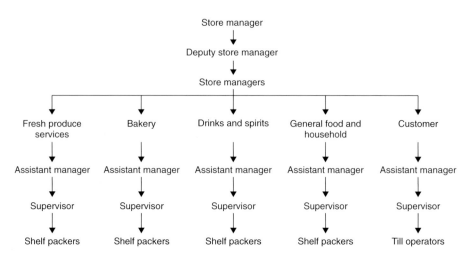

Figure 4.10 Branch structure

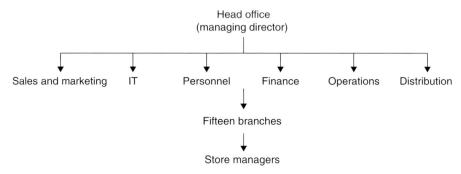

Figure 4.11 Head office structure

All of the functional departments are located at the head office, which has the following implications for each branch:

- All recruitment is done through head office for each of the branches, which means that all the application forms have to be sent either by post or online (if the application was filled in online).
- All stock ordering is done through head office, who have negotiated local supplier contracts for each of the branches.
- All of the promotions, for example 'buy one get one free', and all of the price reductions or special offers are filtered through from sales at head office.
- All salaries are paid via the finance department at head office.
- All deliveries and distribution are made through local suppliers in conjunction with head office instructions.

All of the branches communicate on a regular basis. Branches distribute surplus stock items to other branches if they are running low, to reduce supplier ordering costs.

East Anglia branch

Thomas North has asked for an investigation to take place based on a branch in East Anglia. You have been given some general information about the organizational structure; however, Mr North is keen for you to carry out your investigation at the branch. You have been given four weeks to carry out the analysis and feed back to Mr North.

Fact-finding

Using a variety of fact-finding techniques you managed to collect the following information.

- There are 150 employees at the branch:
    - store manager – Mr Johnston
    - deputy manager – Miss Keyton
    - five store managers, assistant managers and five supervisors
    - fifty full-time and part-time checkout staff
    - forty full-time and part-time shelf stackers
    - ten stock clerks
    - ten trolley personnel
    - three car park attendants
    - twenty cleaners, gardeners, drivers and other store staff.
- Each of the store managers controls their own area, with their own shelf-stackers and stock personnel.
- All stock ordering is batch processed overnight to head office on a daily basis by each of the store managers in consultation with the deputy branch manager.
- All fresh produce is delivered on a daily basis. Non-perishable goods are delivered three times a week by local suppliers.
- All bakery items are baked onsite each morning.

As each department seems to operate on an individual basis, the first part of the investigation for week 1 will be focused on the fresh produce department.

Fresh produce system

Ann Prior and her assistant manager, Mary Granger, manage the fresh produce department. Within the department their supervisor, John Humphries, oversees six display/shelf-stackers and four stock personnel.

After consultation with a range of employees the following account of day-to-day activities has been given.

Each day Ann holds a staff meeting within the department to provide information about new promotions, special discounts or stock display arrangements. Any information regarding new promotions comes through from head office. All information received regarding promotions, etc., is filed in the branch promotions file. If any price adjustments need to be made that day the stock personnel are informed to check the daily stock sheets.

After the meeting the stock personnel liaise with the shelf/display personnel with regard to new stock that needs to go out on to the shopfloor. The information about new stock items and changes to stock items comes from the daily stock sheet. When new items have been put out or stock price adjustments made they are crossed off the daily stock sheet.

Items which have arrived on that day are delivered from the local fresh produce supplier. When the items come in the stock personnel check the daily stock sheet for quantities and authorize the delivery. If items have not arrived or there is an error in the order a stock adjustment sheet is filled in, which is kept in the stock office. At the end of the day John will inform Mary of the stock adjustments. Mary then sends off a top copy of the adjustment sheet to head office and files a copy in the stock cabinet.

Information about stock items running low comes from the daily stock sheet. If an item is low a stock order form is completed. A top copy is sent to head office and a copy is filed in the stock cabinet. Orders should be made five days before the actual requirement for the stock, as head office then processes the information and contacts the local supplier.

In an emergency local supplier information is held by Ann, who can ring direct to get items delivered. This costs the company more money because a bulk order has not been placed. Authorization also has to be given by the operations manager at head office. Information has to be filled in on the computerized stock request form which is e-mailed each day to head office, which then sends back confirmation.

Head office dictates that all documents filled in online also need to have manual counterparts, one which is sent off and another which is filed at the branch.

Problems with the system

- Sometimes the network at head office is down, which means that stock items are not received within five days.
- The promotions are not always appropriate because of a lack of certain stock. Sometimes the branch has a surplus stock which is wasted because they cannot set their own promotions instore.
- The stock cabinet is filled to capacity and because everything is in date order it is difficult to collate information about certain stock items.
- If there is an error in the stock delivery nothing can be supplied until the paperwork has been sent off to head office or authorization has been given, even if the supplier has the stock requirement on his lorry.
- There is too much paperwork.
- There is little communication with other departments.
- Targets that are set by head office cannot always be met owing to the stock ordering problem.
- Some stock items that come in are not barcoded.

Context diagram

The context diagram provides a complete general overview of the system and its relationship with external bodies and entities that are outside the system boundary, as shown in Figure 4.12. The relationship between the system and external bodies is represented through data flows of information between the two.

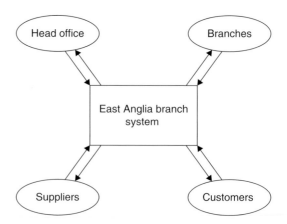

Figure 4.12 Context diagram for TEY Supermarkets

If a context diagram was needed to illustrate just the fresh produce system, fresh produce would appear in the centre of the diagram and other departments within the branch would be added around it as external systems. These would include:

- bakery
- general food and household
- drinks and spirits
- customer services.

Level 1 DFD

Level 1 DFDs provide a general overview of what is happening with the system. The overview will include types of information being passed within the system, documents, storage mechanisms, people, activities and anything that has an impact on processing activities.

The initial system to be examined from TEY Supermarkets is 'fresh produce'. Therefore, the DFD will be based on this specific system.

Ten-step plan in preparing a level 1 DFD

Step 1 Read through the information collected from:
- project brief
- fact-finding investigation
- user catalogues.

Step 2 Sort the information into clear sections, identifying the following:
- who the users are external to the system (sources and recipients of information)
- what documents are used in the system
- what activities take place in the system.

Step 3 Produce a systems information table.

Step 4 Convert external users to external entities.

Step 5 Convert documentation to data stores.

Step 6 Convert activities to processes and identify where the activity takes place and who is involved.

Step 7 Start on a small scale by looking at the inputs and outputs to a single process, using data flows to represent the links of data and information.

Step 8 Position the other processes in the diagram.

Step 9 Connect the remainder of the processes with their attributed inputs and outputs.

Step 10 Check for consistency. Examine initial documentation to ensure that all information has been represented and check back with users or the project sponsor that the diagram is correct.

Step 1 Information collected from the project brief and the fact-finding investigation carried out at TEY Supermarkets

Systems boundary – fresh produce.

Steps 2 and 3 Systems information table

External entities	Data stores (documents)	Processes (activities)
Head office	Promotions file	Daily meeting
Supplier	Daily stock sheet	Put out and adjust stock
	Stock adjustment sheet	Check deliveries
	Stock cabinet	Order stock
	Stock order forms	
	Stock request forms	

Activity 4.11

Identify the system that you use for getting up in the morning and getting ready for work or college. Identify the activities that take place, who is involved in your system and any documents that you may use to get you to college, ready for your first lesson.

Steps 4, 5 and 6 Conversion of information into DFD tools

Sample external entity   Sample data store   Sample process

Step 7 Single process DFD

Steps 8 and 9 Connection of the remainder processes

Level 1 DFD for the fresh produce department at TEY Supermarket:

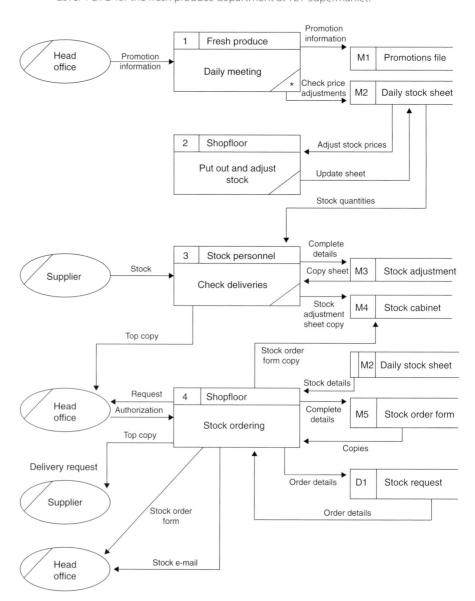

Step 10 Check information

Information is accurate and complete, checked against information provided by the personnel within the fresh produce department

The ten-step plan is a guide to preparing DFDs. Different people will use their own methods. The benefit of the plan is that you are constantly re-examining the information, and understanding the system is half the battle when preparing DFDs.

Level 2 DFD

A level 2 DFD provides a more detailed view of a specific process that has been represented at level 1.

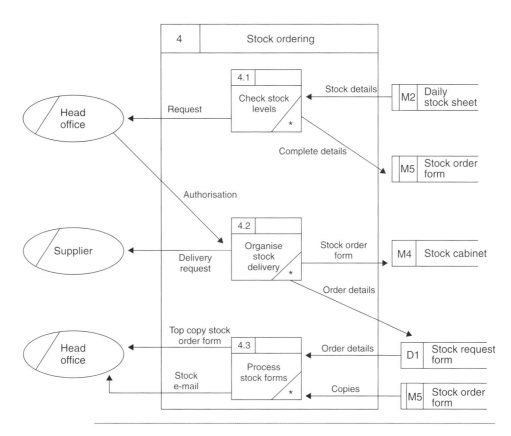

Activity 4.12

1. Within the class, identify whether anybody has a part-time job or hobby at the weekends or in the evenings. Ask them to talk through and explain what they do as part of their job or hobby. Record all of the information with regard to:
   - what tasks they do
   - who they communicate with
   - what documents they use
   - where information comes from and what happens to the information.
2. From the information given produce a context diagram, level 1 and level 2 data flow diagram (if appropriate) for the system.
3. One person should then draw the diagram on the board so that everybody can contribute to the representation of the system; at each stage checks should be made with the user to ensure that all information is correct.

CHAPTER 4

Documentation

Along with the diagrams, the second element of data flow modelling is the documentation that accompanies them. The documentation is used to support and clarify areas of the diagram that may be ambiguous.

Documentation that can be used in data flow modelling includes:

- elementary process descriptions (EPDs)
- external entity descriptions
- input/output descriptions.

The different levels of DFD contain different types of processes. Processes at level 1 may be broken down further to level 2. If a process cannot be broken down any further, this is indicated by putting an asterisk in the bottom right-hand corner of the process box (Figure 4.13). This notation means that there is no further composition of the process. Each elementary process will then have an associated EPD. The descriptions identify the activities or operations that take place, as shown in Table 4.3.

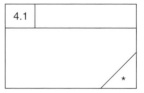

Figure 4.13 Example of an elementary process

Table 4.3 Example of an external entity description for the head office at TEY Supermarkets

Entity name	Description
Head office	Central functional branch that provides all of the information regarding: • promotions and special discounts • daily price adjustments • stock delivery details • authorization of emergency supplier stock deliveries All top copies of documentation are forwarded to head office each day

External entity descriptions detail the status of the external entity in terms of identification of the role and responsibilities it has as part of the system.

Input/output provide textual descriptions of the data flows that extend across the system boundary providing links to external entities.

Data flow modelling provides the analyst with a visual tool from which accurate representations of the system can be drawn. DFDs assist in identifying the activities and the data that is used within the system. and the textual descriptions provide support and clarity.

Activity 4.13

Using the information provided by TEY Supermarkets and by examining the data flow diagram of the system, address the following:

1. Identify the problems at TEY Supermarkets.
2. Identify any problems that can be seen from the data flow diagram.
3. Provide a range of solutions to these problems.

Data dictionaries – these provide information about data stored within a database. A data dictionary does not contain any data, just information to manage it. It contains information about the fields, size and data types.

Entity relationship modeling

Another tool that is used to identify and represent the activities of the system is logical data modelling. This tool provides a detailed graphical representation of the information used within the system and identifies the relationships that exist between data items. Similarly to data flow modelling, logical data modelling uses a set of tools and associated textual descriptions.

The diagrammatic aspects of entity relationship modelling are referred to as entity relationship diagrams (ERDs). These diagrams have four main components (Figure 4.14).

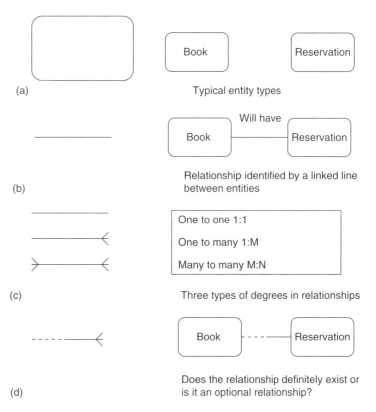

Figure 4.14 Entity relationship diagram notations: (a) entities; (b) relationships; (c) degree; (d) optionality

- entities
- relationships
- degree
- optionality.

Entities

Entities provide the source, recipient and storage mechanism for information that is held on the system. Typical entities are shown for three systems, as follows.

Library system
Entities:
- book
- lender
- reservation
- issue
- edition

Hotel system
Entities:
- booking
- guest
- room
- tab
- enquiry

Airline system
Entities:
- flight
- ticket
- seat
- booking
- destination

Each entity will have a set of attributes that make up the information occurrences, for example:

Entity:
- book

Attributes:
- ISBN
- title
- author
- publisher
- publication date

Each set of attributes within that entity should have a unique field that provides easy identification to the entity type. In the case of the entity type 'book' the unique key field is that of 'ISBN'. The unique field or 'primary key' will ensure that although two books may have the same title or author, no two books will have the same ISBN.

Relationships

To illustrate how information is used within the system, entities need to be linked together to form a relationship. The relationship between two entities could be misinterpreted; therefore, labels are attached at the beginning and at the end of the relationship link to inform parties exactly

Primary key – a unique key field, used to provide direct access to the entity, a unique identifier.

what the nature of the relationship is. For example, if you had two entities linked as illustrated in Figure 4.15, the nature of the relationship could be any of the following:

- An author can write a book, therefore the book belongs to an author.
- An author can refer to a book, therefore the book is in reference by an author.
- An author can buy a book, therefore the book is bought by an author.
- An author can review a book, therefore the book is reviewed by an author.

Figure 4.15 Example of entity relationships

The actual relationship that exists in this scenario is that an author reviews a book, therefore the book has been reviewed by an author.

Degree

There are three possible degrees of any entity relationship.

- One to one (1:1) denotes that only one occurrence of each entity is used by the adjoining entity, for example a single author writes a single book:

- One to many (1:M) denotes that a single occurrence of one entity is linked to more than one occurrence of the adjoining entity, for example a single author writes a number of books:

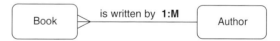

- Many to many (M:N) denotes that many occurrences of one entity are linked to more than one occurrence of the adjoining entity, for example an author can write a number of books and books can have more than one author:

- Although M:N relationships are common, the notation for linking two entities directly is adjusted and a link entity is used to connect the two. For example, a customer can make a number of bookings and each of those bookings is made:

or a customer can make a number of enquiries which lead to a booking and bookings result from a number of enquiries made by customers:

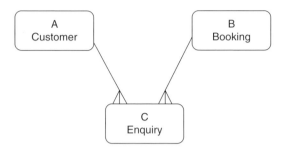

Optionality

There are two status types given to a relationship: those that definitely happen or exist, and those that may happen or exist, this second status being referred to as 'optional'.

A dashed rather than a solid link denotes optionality in a relationship:

In this scenario a customer may or may not decide to make a booking. If they do, the booking will definitely belong to/be made by a customer.

Activity 4.14

1. What are the three types of relationship degrees? Give an example of each.
2. Provide sample diagrams to meet the following criteria:
   - A customer buys a concert ticket.
   - A guest makes a number of complaints about a hotel room.
   - A doctor examines a patient.
   - A receipt is given to a customer as a result of a purchase.
3. What are data items called that belong to an entity?
4. How could you easily identify an entity?

Activity 4.15

Using the information given below, taken from TEY Supermarkets, design an entity relationship diagram, clearly labelling all relationships.

'Items which have arrived on that day are delivered from the local fresh produce supplier. When the items come in the stock personnel check the daily stock sheet for quantities and authorize the delivery. If items have not arrived or there is an error in the order a stock adjustment sheet is filled in, which is kept in the stock office. At the end of the day John will inform Mary of the stock adjustments. Mary then sends off a top copy of the adjustment sheet to head office and files a copy in the stock cabinet.

'Information about stock items running low comes from the daily stock sheet. If an item is low a stock order form is completed. A top copy is sent to head office and a copy is filed in the stock cabinet. Orders should be made five days before the actual requirement for the stock, as head office then processes the information and contacts the local supplier. In an emergency local supplier information is held by Ann, who can ring direct to get items delivered. This costs the company more money because a bulk order has not been placed. Authorization also has to be given by the operations manager at head office. Information has to be filled in on the computerized stock request.'

Entity relationship modelling documentation

The second component of entity relationship modelling is the documentation that supports the ERD. The documentation that is used includes:

- entity descriptions
- attribute lists

Every entity should have an associated entity description, which details items such as:

- entity name and description
- attributes
- relationship types and links.

Every entity has a set of attributes. If a large system is being investigated a number of entities and their associated attributes will need to be defined, therefore an attribute list can be prepared.

Attribute lists identify all of the attributes and a description of the attributes. The primary key attribute, which is normally made up of numerical data, e.g. supplier number, National Insurance number, examination number, is referred to first, followed by the remainder of the attribute items.

Activity 4.16

1. What should an entity description include?
2. Why should entity relationships be labelled?
3. Provide sample attribute lists for the following entities:
   - appointment
   - invoice
   - menu
   - supplier.
4. What would the primary key be in each case?

Decision tables

Decision tables provide a simple way of displaying certain actions that occur under certain conditions. Decision tables provide a visual representation of this by clearly defining sections devoted to stubs and entries (Table 4.4).

The advantages of preparing a decision table are that all combinations of conditions will be considered and that there is a clear overview of what conditions have been met or not met. The standard layout also ensures that information is clearly understood and can be used by a number of end-users.

CHAPTER 4

Table 4.4 Decision table

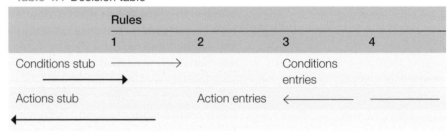

	Rules			
	1	2	3	4
Conditions stub			Conditions entries	
Actions stub		Action entries		

Structured English – uses a series of statements to describe the processes within a system. Statements that are used include a range of keywords such as IF, THEN, DO and ELSE.

Flowcharts

Flowcharting is a traditional design tool that is still used today. System flowcharts detail the flow of information through an entire information system using specific symbols to characterize sequences and processes (Figure 4.16).

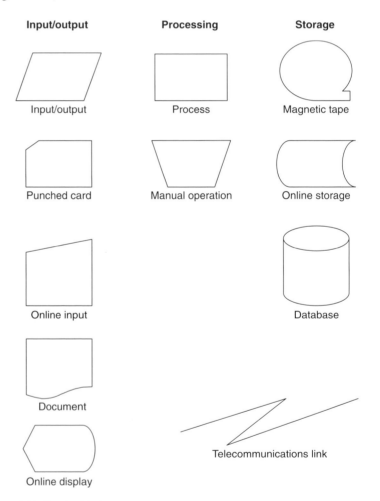

Figure 4.16 Flowchart tools

Constraints

When you design a system there are many factors and influences that can affect the type of system, cost, resources used, implementation and testing strategies. There are also certain conditions within which you have to operate; these are also referred to as constraints.

System design constraints may revolve around costs and budgets, therefore the perfect design may have to be modified to fit within a certain budget. Constraints can also be based around organizational policies, therefore the system may have to conform to certain organizational policies and procedures. Having to work within a given time-frame can impose huge constraints on a system design. Further, having to work with existing legacy systems that may not have compatible hardware or software, or a system that is not capable of interacting across different platforms, can be detrimental to any design proposal.

Be able to design a test plan

Once a system has been designed and implemented there is a need to test it. Testing is a crucial part of any system and can be carried out incrementally as each element is completed or goes 'live' or at the end of the implementation process. There are different testing strategies to test different components of a system, for example the hardware or the software.

Testing strategies

Testing is paramount to the development and implementation of software. The stages of software testing will vary depending on which phase is being addressed. For example, during the prototyping phase testing may be more ad hoc as the focus is to identify any features that are missing or define different ways of performing a task or function. During the implementation phase, however, testing becomes more structured so as to identify as many faults as possible. Testing strategies include white-box, black-box and V-model testing.

White-box testing

This technique is used to test the internal structures and code of a programme or module and each test is usually based on the logic of the code. The objective is to test each path through the code and prove that it is executed correctly. This kind of testing is sometimes called statement coverage because it requires the tester to check the code rigorously.

For example, where an IF statement has a condition that can evaluate as true or false, the test case needs to cover both possibilities.

Black-box testing

This technique tests the inputs and outputs of a block of code, procedure or complete programme, but without regard to the internal logic or code; hence black-box: it is only interested in what goes in and what comes out. This form of testing can include user actions and the passing in and out of data.

V-model testing

Throughout the software life cycle a testing model can be applied that identifies the various aspects of testing throughout analysis and design to test the requirements specification (user acceptance test) and the detailed design (unit testing). This model is known as the V-model (Figure 4.17).

CHAPTER 4

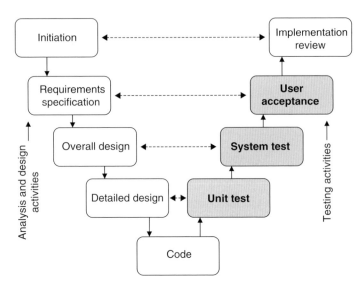

Figure 4.17 V-model of testing in relation to analysis and design

Although software may go through a number of testing stages, it may still fail to work or initiate successfully when implemented as part of an IT system. The reasons for this may include:

- incompatibility of coding, programmes and data across platforms
- level and expertise of the end-user
- problems with the systems hardware – not fast enough, inadequate memory, etc.

Testing can be broken down into a number of levels and stages, structured in the form of a hierarchy (Figure 4.18).

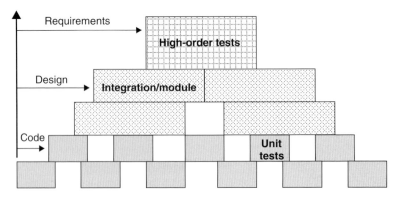

Figure 4.18 Testing hierarchy

High-order tests

High-order testing comprises a number of areas:

- validation tests (alpha and beta tests)
- system tests
- other specialized tests (performance, security, etc.).

Alpha and beta tests (Figure 4.19) take place before a packaged software release. The purpose of alpha tests is to identify any bugs or major problems that may affect software functionality or usability in the early build phase. This type of testing is usually carried out inhouse by staff members.

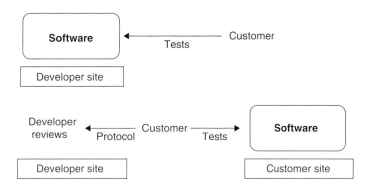

Figure 4.19 Overview of alpha and beta tests

Beta tests occur after the alpha tests to identify any bugs in the software before it is released to customers.

Integration and module tests can be broken down into a number of possibilities, including:

- big bang
- top–down
- bottom–up
- regression testing.

The unit testing environment is shown in Figure 4.20. Unit testing tests individual modules to ensure that they function correctly for any given input or inputs.

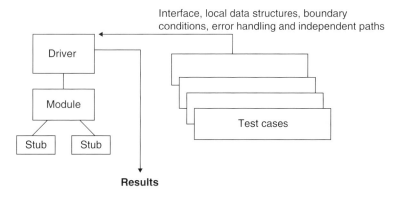

Figure 4.20 Unit testing environment

The function of testing

Testing is vital to any system design. Testing can be carried out at any stage as an incremental process or at the end of implementation. Whenever testing is carried out the function remains the same: the need to instil quality and checks that the user requirements have been met, that the system proposal meets the required specification and that it is accepted is paramount.

CHAPTER 4

On a more practical level, testing will ensure that components operate effectively, uniformly and consistently at all levels.

Test plan

The test plan provides documentary evidence that testing has been carried out and should take into account the following points:

- Version control – what test number/version are you up to?
- What has been tested?
- At what stage in the development?
- The purpose of the test.
- The results of the test.
- Comments – did the test run as expected?

Each time a test is carried out the test plan should be updated and used as a working document that can be integrated into the final evaluation. An example test plan is presented in Table 4.5

Table 4.5 Sample test plan

Test ID	Description	Expected result	Actual result	Pass/fail
	Application start/exit			
1	Application starts correctly	Display application menu screen	Menu screen loads	P
2	All menu screen options/ buttons function correctly	Options cause correct screens to be displayed	All options OK	P
3	Application closes down correctly	Application closes	Application fails to close	F

Questions and review

1. Can you identify and describe two development life cycle models?
2. What generic stages might you find across different life cycle models?
3. How would you describe a developmental methodology?
4. Data flow diagrams are an important modelling tool that is used within systems analysis. Why are they important and what do they do?
5. What are the key drivers for systems analysis and design?
6. There are a number of benefits associated with effective systems analysis procedures, what are they?
7. There are a range of investigative techniques that can be used in a systems analysis investigation. Identify and describe the benefits of using three of these techniques.
8. What are the main components/elements of a requirements specification?
9. There is a range of documentation that can be produced and used when creating a system. Identify three different documents that could be used and describe their role within the system design stages.
10. What constraints should be considered when creating a systems design?
11. Identify a range of test strategies that could be used when designing and implementing a new system.
12. Why is it important to have a test plan?
13. What, in your opinion, is the purpose of testing?

Assessment activities

Grading criteria	Content	Suggested activity
Pass		
P1	Describe the key drivers for initiating a particular systems analysis activity and explain the particular potential benefits.	Produce a 'feasibility report' that will incorporate a number of grading criteria to include P1, P4, P5, M1 and M2. The initial report would incorporate P1 in terms of describing the key drivers for initiating a particular systems analysis activity and explaining the particular potential benefits.
P2	Describe two different development life cycle models and their suitability for particular situations.	Conduct research into different life cycle models and describe two of them and their suitability for particular situations. This could be presented as a written short summary with diagrams of the chosen life cycles.
P3	Describe the features, advantages and disadvantages of a development methodology.	You could include a description of features, advantages and disadvantages of a development methodology in your summary.
P4	Undertake an investigation and document the specification requirements.	Your tutor might give you a scenario that would involve you conducting an investigation. The scenario could for example be based on a travel agency booking system or a library reservation system. It might be possible to use a real-life case study based on an actual organisation such as a college department or system, or your own place of work. An additional report section should be included to identify what steps have been taken in the investigation, methods used for fact-finding and evidence of using modelling tools. You should also document the specification requirements within the report.
P5	Design and document a new system.	Students should provide outlines, draft and actual, of their system design(s) to include all of the necessary documentation to support it.
P6	Design a test plan.	Design a test plan for their new system design.
Merit		
M1	Describe alternative solutions to a given requirements specification and explain in what circumstances such alternative solutions would be chosen.	The descriptions for alternative solutions can be provided within the feasibility report. This criteria can easily be linked with P4, offering alternative solutions to the given requirements specification and identifying the circumstances in which they may be chosen, for example: greater budget, more time, access to more advanced technology or resources etc.
M2	Take an independent role in planning the production of a requirements specification.	Evidence for this can also be incorporated within the report. You could state how you have planned the production of your requirements specification.
M3	Explain the choice of a particular test strategy.	In conjunction with P6, explain/justify why you have used a particular test strategy for your design. This written explanation can complement the test plan produced in P6.
Distinction		
D1	Judge the effectiveness of a systems analysis process in ensuring that the design is fit for purpose and recommend potential improvements.	Give an evaluative or reflective account/presentation of the effectiveness of the systems analysis process in ensuring that the design is fit for purpose and also how it could have been improved.
D2	Explain the potential risks in undertaking a systems analysis and design activity and for each one offer ways of reducing or eliminating that risk.	Embedded within the presentation for D1, further slides could be included that explain the risks of undertaking a systems analysis design activity and suggesting ways in which the risks could be reduced or eliminated.

Courtesy of iStockphoto, duckycards, Image# 396641

Communications technologies covers a range of IT devices and disciplines. Over the years communications technologies have changed to meet the demands of society, with the emphasis on portable and wireless devices that can support the mobility and flexibility of users in the 21st century.

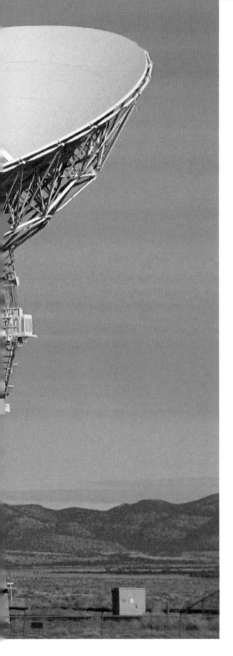

Communications Technologies

The methods and tools associated with communications technology have changed over recent years to reflect the demands of users and the need for more portable and wireless devices from which to transmit and receive communications.

This chapter will explore the way in which communications technologies have developed and are used in both a professional work-based and social environment.

The chapter will provide an overview of communications technology devices, transmission rates, connectivity and associated protocols. More specifically, the chapter will focus on the following learning outcomes:

- Know the main elements of a data communications system.
- Understand the communication principles of computer networks.
- Understand transmission protocols and models.
- Understand Internet communications.

This chapter will provide you with information and a range of activities that can support you with your assessment(s) for this unit.

Know the main elements of a data communication system

Data communication systems cover a range of elements, theories, media and devices which will be explored in the following section.

Communication devices

Communication devices can be categorized in terms of portability, for example wired devices such as data terminal equipment (DTE) or data circuit-terminating equipment (DCE). Communication devices can also be wireless, for example 3G cellular phones, wireless personal data assistants (PDAs) or wireless laptops.

DTE – an 'end instrument' (a piece of equipment connected to the wires at the end of a telecommunications link) that can convert user information into signals ready for transmission purposes. DTE can also reconvert any received signals into user information.

DCE – can also be referred to as 'data communications equipment' or 'data carrier equipment'. The DCE is the device that is positioned between the DTE and a transmission circuit.

PDA – a handheld device (Figure 5.1), originally used as a personal organizer; however, it has developed into a device that has many functions, including the integration of games and applications and the ability to access e-mail and the Internet through Wi-Fi connectivity.

Figure 5.1 PDA: Acer n50 (http://www.acer.co.uk)

Activity 5.1

PDAs are very popular for communicating on the move, in the absence of a desktop or laptop device. Produce a one-page information leaflet on PDAs to include a range of the following items:

- types of PDA
- prices
- features
- benefits
- users.

Signal theory

Successful electronic data transmission requires the use of a signalling method that is recognized by both the 'transmitter' and the 'receiver' and the medium that connects them. Data can be transmitted by either analogue or digital methods (Figure 5.2).

Figure 5.2 Example of analogue and digital transmission

Analogue transmission is used to carry voice, data or fax, and has a limited bandwidth. Using a waveform, messages are carried by a cable such as a telephone line. Analogue information is transmitted by modulating a continuous transmission signal. A modem, for example, modulates data that is received over a telephone line in analogue to a digital format that can be understood by a computer and demodulates received signals to retrieve data.

When data is transmitted using analogue methods, a certain amount of noise (data without any meaning) enters into the signal.

A digital signal is not a continuous wave like analogue; it consists of binary data sent in '1s' and '0s', 'on' or 'off' that is interpreted by computers. The Internet is a network of digital signals, as are most mobile phone technologies.

Data transmission can also be characterized by its timing in terms of sending data in synchronous blocks or irregular intervals, asynchronously.

Synchronous and asynchronous transmission

Synchronous transmission sends data in blocks or frames, the frames varying in size between 1000 and 4096 bytes. Information that is contained within a frame is shown in Figure 5.3. Synchronous transmission is more efficient than asynchronous communication because the number of non-data bits is smaller since asynchronous transmission has an additional 25 per cent more bits added for the control element.

Flag	Address	Control	Data	CRC	Flag

Figure 5.3 Synchronous data frame

Synchronous transmission requires more sophisticated resources to support it, therefore it is generally used within a networking environment, whereas asynchronous is used mainly to connect simple devices.

Asynchronous transmission is the more common of the two and data is sent as a series of 1s and 0s. If no data is being sent the line is quiet. It is activated only when a byte is going to be sent, when a start bit will begin the transmission. A problem with asynchronous transmission is that each byte has more added, therefore more bits are sent than are required by the data alone.

Error detection and correction

When data is being transmitted it is important that it retains its integrity, especially if it is travelling over noisy channels or less reliable storage media. Error detection/correction is important because it helps to maintain the integrity of the data by detecting any errors caused by noise or other impairments during transmission and has the ability to reconstruct the original error-free data.

Various error detection schemes exist:

- parity schemes – can detect an odd number of bit errors
- polarity schemes – more effective in detecting and correcting errors within the physical layer of the open system interconnection (OSI) model
- repetition schemes – a simple way of breaking down data into blocks of bits and sending each block a predetermined number of times
- cyclic redundancy checks – a more complex error detection method that makes use of the properties of finite fields and polynomials over such fields.

Other schemes include the use of checksum and hamming distance-based checks.

Error correction methods provide further support by correcting any errors identified. A number of methods can be used to carry out this function, including:

- Automatic repeat request uses acknowledgement messages and timeouts to send a message to the receiver to indicate that data frame has been correctly received. If the acknowledgement has not been received before the timeout period then the frame is retransmitted until it is received in the correct format or the error continues beyond a set number of attempts.
- Error correcting code (ECC) provides a code that is written within the rules of construction for each data signal. If a signal is received that does not comply with these rules of construction then errors can be detected and corrected.

Effect of bandwidth limitation and noise

Bandwidth can be used to refer to the speed at which data can be downloaded, which is measured in bytes per second. A 4 Kbyte link would be classed as a low bandwidth, whereas 100 Mbytes would be considered to be high.

When a signal is transmitted through a communication channel, the levels can drop off at some point owing to the physical nature of the channel.

Some communication channels can transmit high-power signals; however, some cannot and suffer from noise – random signals caused by outside interference. The cause of this interference could include:

- radio frequencies
- other electrical devices
- random activity of the electrons in the wires.

Activity 5.2

Signal theory examines a range of data signalling methods. It also looks at how data can be represented, transmission media, and error correction and detection. There are other areas that should be examined within the topic of signal theory to see how these can impact on signalling methods and transmission.

Complete the table by identifying how:

- bandwidth
- channel type
- other issues such as data compression

can affect signal theory.

Influence on signal theory methods and transmission	How it can influence signal theory methods and transmission
Effect of bandwidth (limitation and noise)	
Channel type, e.g. telephone, high-frequency (HF) radio, microwave, satellite	
Other issues, e.g. data compression, bandwidth	

Data elements

There are various data elements within data communication systems, all of which perform a variety of tasks and functions. These elements include:

- checksum, e.g. cyclic redundancy check (CRC)
- packets and frames
- datagrams.

A checksum is a form of redundancy check; it protects the integrity of the data by detecting errors in the data that is sent. CRC is a more sophisticated form of checking that addresses weaknesses in terms of considering the value of each byte and its position during transmission.

Data that travels through networks does so in small groups of bytes called packets, and the whole process is controlled by a set of rules called protocols. As packets find their way around the network they should have information attached that provides the destination address and source data, in conjunction with other information. When packets are wrapped up with this data the result is called a frame.

A datagram can also be referred to as a packet; it is a small piece of data that is self-contained. Datagrams have a source and a destination but no connection elements or relationships with anything that preceded or comes after them.

Methods of electronic communication

Duplex communication means that a signal can travel in both directions between connected parties/media/technologies (Figure 5.4). A full-duplex communication system facilitates this two-way communication, an example being a telephone where two people can speak simultaneously. Half-duplex communication only allows transmission in one direction at a time, e.g. a walkie-talkie where one person speaks and then the other person speaks.

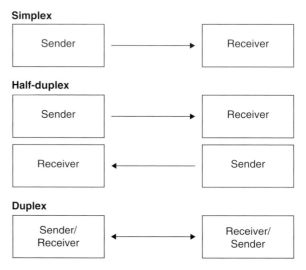

Figure 5.4 Simplex, half-duplex and duplex transmissions

In terms of using a modem, some modems have a switch that allows you to select between half- and full-duplex modes. In half-duplex mode each character transmitted is immediately displayed on your screen. In full-duplex mode transmitted data is not displayed until it has been received and returned.

In addition to two-way transmission, there is a one-way transmission known as 'simplex'.

Electronic communication can also be found in the form of parallel and serial interfaces such as universal serial bus (USB), FireWire, RS-232 and a range of other technologies such as infrared, Bluetooth® and Wi-Fi.

Transmission media

Fibreoptic cables are used in telephone networks, and play a significant role in long-distance telephone routes which span over 1500 km; they are also used in rural trunking. Optical cables are also used in local area networks and television networks.

Coaxial cables are more commonly used for carrying television signals; however, in recent years they have also been used in telephone networks, with a single coax cable being capable of carrying over 10 000 voice channels simultaneously.

Twisted pair is most commonly used in local area networks and telephone networking and is usually the cheapest transmission medium available. A twisted pair is made of two copper wires twisted together. There are two types:

- unshielded twisted pair (UTP)
- shielded twisted pair (STP).

Activity 5.3

1. Carry out research and produce an information sheet based on at least four different communication methods/transmission media. These could include:
   - Bluetooth
   - Wi-Fi
   - coaxial
   - optical fibre
   - unshielded twisted pair (UTP)
   - shielded twisted pair (STP)
   - FireWire
   - universal serial bus (USB).
2. Group activity – students should draw up a table of different transmission media (as listed) and compare the different features and benefits of each:
   - infrared
   - radio
   - satellite
   - microwave.

Understand the communication principles of computer networks

This section will help you in your understanding of computer networks by examining their features, components and interconnection devices.

Computer networks provide an interconnected system or systems that aid communications and data transmission across a number of physical locations. Computer networks can be quite complex, as they are designed around a certain topology, provide a number of services, and support a range of hardware and software, some of which is exclusive to a networked environment or specific platform.

Features of networks

A networked system can bring many benefits to an organization:

- sharing of data and information and the dissemination of good practice
- increased efficiency
- sharing of resources, e.g. printers and scanners
- reduced information transfer time
- reduced costs.

As a result of these benefits many organizations opt for a networked solution despite the initial financial outlay, setup costs, possible disruption to employees, and the need to train and update.

Networks vary in size and complexity. Some are used in a single department or office, while others extend across local, national or international branches. Networks vary in structure, to accommodate the need to exchange information across short or wide geographical areas. These structures include:

- local area networks (LANs)
- metropolitan area networks (MANs)
- wide area networks (WANs) – long haulage networks (LHNs)
- value added networks (VANs).

Local area networks

These consist of computers that are located physically close to each other, within the same department or branch. A typical structure includes a set of computers and peripherals linked as individual nodes. Each node, for example a computer and shared peripheral, is directly connected by cables that serve as a pathway for transferring data between machines.

Metropolitan area networks

These are more efficient than a LAN and use fibreoptic cables, allowing more information and a higher complexity of information. The range of a MAN is also greater than a LAN, allowing business to expand around a country; however, this can prove to be expensive because of the fibreoptic cabling.

Wide area networks – long haulage networks

These are networks that extend over a larger geographical distance, from city to city within the same country or across countries and even continents. WANs transfer data between LANs on a backbone system using digital, satellite or microwave technology.

A WAN will connect different servers at a site. When this connection is from a personal computer (PC) on one site to a server on another it is referred to as being 'remote'. If this coverage is international it is referred to as being an 'enterprise-wide network'.

Value-added network

This type of network is a data network that has all the benefits of a WAN but with vastly reduced costs. The cost of setting up and maintaining this type of network is reduced because the service provider rents out the network to different companies, rather than an organization having sole ownership or a 'point-to-point' private line.

Networks can be used to support a range of applications within an organization, the selection of a particular network depending on:

- the application/use
- the number of users requiring access
- physical resources
- the scope of the network, within a room, department, across departments or branches.

If, for example, a network was required to link a few computers within the same department to enable the sharing of certain resources a LAN may be installed. If, however, a network was required to link branches and supplier sources across the country a WAN may be installed, with the server located at the head office providing remote access to users connecting at individual branches.

Developments in technology have given way to new product developments in terms of communication devices; one of these is the wireless LAN (WLAN), offering a more flexible approached to wired networks by using radiowaves as the medium of transmission.

Network topologies

Network topologies (topology referring to the layout of connected devices on a network) include:

- bus
- ring
- star
- tree
- mesh.

A bus topology is based on a single cable which forms the backbone to the structure with devices that are attached to the cable through an interface connector (Figure 5.5). When a device needs to communicate with another device on the network a message is broadcast. Although only the intended receiving device can accept and process the message, a disadvantage of this topology is the fact that other devices on the network can also see the message.

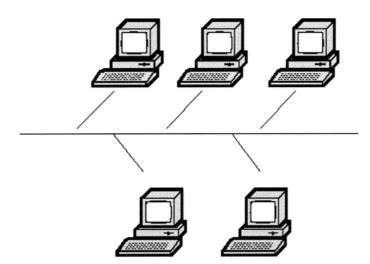

Figure 5.5 Bus topology

In a ring topology every device has two neighbours for communication purposes (Figure 5.6). Communication travels in one direction only, either clockwise or anticlockwise. One disadvantage of this type of topology is that a failure in the cable can disable the entire network.

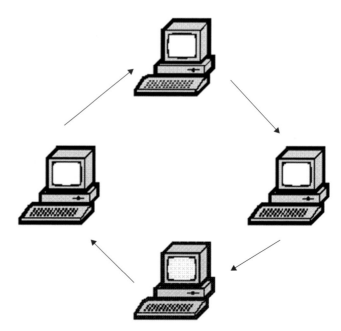

Figure 5.6 Ring topology

Star topologies are based around a central connection point, referred to as a 'hub' or a 'switch' (Figure 5.7). A series of cables is used in this type of network, giving it the advantage that if one cable fails only one computer will go down and the remainder of the network will remain active. If the hub fails, however, the entire network will fail.

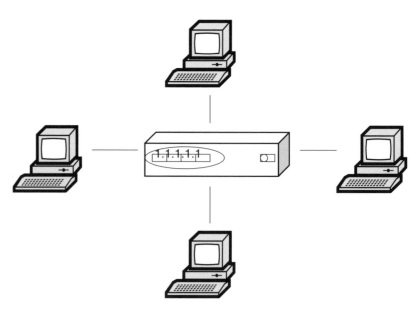

Figure 5.7 Star topology

Tree topologies integrate a number of star topologies together onto a bus, with the hub devices connecting directly to the tree bus, each hub acting as the 'root' of the tree device.

A mesh topology introduces the concept of 'routes', where several messages can be sent on the network via several possible pathways from source to destination.

Activity 5.4

1. Identify the benefits that a networked system can bring to an organization.
2. Network structures can be described as being LANs, MANs, WANs and VANs. Compare and contrast each of these structures in terms of their capabilities and benefits to an organization.
3. What is meant by a network 'topology'? Provide a definition of at least three types.
4. Define the key components in a client/server system and give examples of different server types.

Networks can be complex because they are reliant on different components, hardware, software, media and services all communicating together to provide the necessary data transmission provision. Networking hardware comprises the physical system, PC, peripheral items, connectors, wires, hubs, switches and cabling, etc. Network software covers elements such as the network operating system and connection software. Network services can be:

- packet switched
- integrated services digital network (ISDN)
- multiplexed
- asynchronous transfer mode (ATM)
- wireless applications protocol (WAP)
- broadband.

Activity 5.5

Carry out research into the following network services and complete the table of benefits and drawbacks for each.

Service	Benefits	Drawbacks
Packet switched		
ISDN		
Multiplexed		
ATM		
WAP		
Broadband		

Access methods

In networking you need to access a resource in order to use it. Access methods are the rules that define how a computer puts data onto a computer cable and takes it off. Once data is flowing across the network,

CHAPTER 5

the access method regulates this flow of traffic. Resources can be accessed by:

- carrier-sense multiple access:
  – with collision detection (CD)
  – with collision avoidance (CA)
- token passing
- demand priority.

Carrier-sense multiple access with collision detection checks the cable for each computer on the network for 'network traffic'. As soon as a computer senses that the cable is traffic free it can send its data. Once this has been transmitted, no other computer can send data until the original data has reached its destination and the cable then becomes free again.

A data collision will occur if two or more computers decide to send data simultaneously. When this happens the computers that are attempting this data transmission will need to stop and then after a certain period try to retransmit. The period of waiting is determined by each computer involved, thus reducing the chance of another collision caused by simultaneous data transmission.

Carrier-sense multiple access with collision avoidance is a method whereby each computer signals when it is about to transmit data, prior to sending. With this method other computers will know that a collision may happen and therefore will wait until the cable is free from traffic.

Token passing uses a special packet called a 'token' which circulates around a cable ring passing from computer to computer. In order for a computer to send data across the network it must wait for a free token. When the token is detected by the computer and it has control of it, it will be able to send the data.

This access method, known as demand priority, is designed for the 100 Mbps Ethernet standard known as 100 VG-AnyLAN. It is based on the notion that repeaters and end nodes are the two components that make up all 100 VG-AnyLAN networks. The repeaters manage the network access by doing round-robin searches for requests to send from all nodes on the network. The repeater, or hub, is responsible for noting all links, addresses and end nodes and checking that they are all working accordingly.

End node – can be a computer, router, bridge or switch.

Network components

Networks consist of a number of components, such as servers, workstations and network cards.

Servers

All of the machines on the Internet are either servers or clients. The machines that provide services to other machines are servers, while the machines that are used to connect to those services are clients. Servers can be categorized into the following:

- web servers
- e-mail servers
- file transfer protocol (FTP) servers
- newsgroup servers.

When you connect to a website to read a page, you are accessing that site's web server. The server machine finds the page you requested and sends it to you. Clients who come to a server machine do so with a specific intent, so clients direct their requests to a specific software server running on the server machine.

The server is a more powerful computer that stores the application and the data that is shared by users. Servers effectively circulate the information around the network and together with the network operating system perform a number of functions, as shown in Figure 5.8.

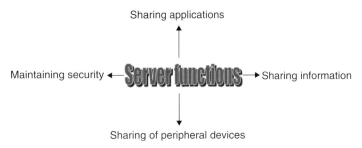

Figure 5.8 Server functions

Applications and data can be managed more effectively when they are managed by a server. Auditing functions can also be undertaken more easily to ensure that data is being kept secure.

Within larger networks there may be servers that are dedicated to a specific resource or function; for example:

- print server
- file server
- mail server

and a number of servers that are reserved for data storage.

Workstations

These are higher performance machines than normal PCs. They are high-end machines designed specifically for technical applications.

Network card

A network card, sometimes referred to as a network interface card (NIC), plays the very important role of connecting your cable modem and your computer together. It will enable you to interface with a network through either wires or wireless technologies.

The network card allows data to be transferred from your computer to another computer or device. Examples of network cards include Ethernet, wireless and token ring.

Interconnection devices

Hubs

A hub can join multiple computers or other network devices together to form a single network segment where all computers can communicate directly with each other. Hubs operate at the physical layer in the OSI model. They do not support any sophisticated networking features, they cannot read any passing data and they are not aware of their source or destination. There are three main types of hub:

- passive
- active
- intelligent.

Switches

A switch can connect Ethernet or other types of packet switched network segments together to form a diverse network that operates within the data link layer of the OSI seven-layer model and sometimes the network layer.

Routers

Routers determine where to send the information from one computer to another. They are specialized computers that send the messages quickly to their destinations along thousands of pathways.

A router serves two purposes. First, it ensures that information does not go astray, which is crucial for keeping large volumes of data from clogging up connections; and secondly, it ensures that the information does indeed make it to the intended destination. In performing these two roles a router is invaluable. It joins networks together, passing information from one to another, and also protects the networks from each other. It prevents the traffic from one unnecessarily spilling over to another.

Regardless of how many networks are attached, the basic operation and function of the router remain the same. Since the Internet is one huge network made up of tens of thousands of smaller networks, the use of routers is an absolute necessity. In order to handle all the users of even a large private network, millions and millions of traffic packets must be sent at the same time.

Some of the largest routers are made by Cisco Systems Inc. Cisco's 'Gigabit Switch Router 12000 series' is a typical router system that is used on the backbone of the Internet (Figure 5.9). These routers use the same sort of design as some of the most powerful supercomputers in the world, a design that links many different processors together with a series of extremely fast switches.

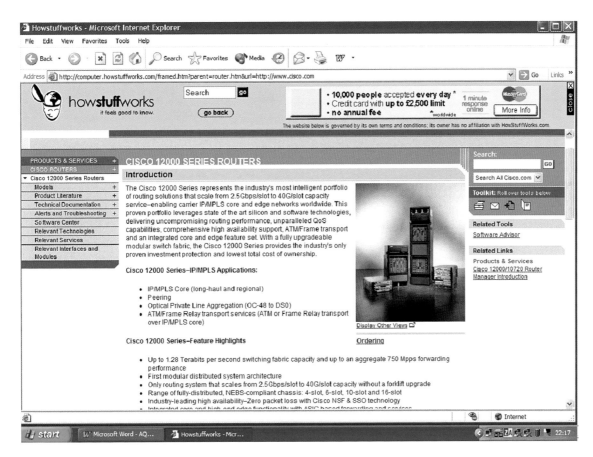

Figure 5.9 Cisco 12000 series router
http://www.howstuffworks.com

The 12000 series uses 200 MHz MIPS R5000 processors, the same type of processor used in the workstations that generate much of the computer animation and special effects used in movies. The largest model in the 12000 series, the 12016, uses a series of switches that can handle up to 320 billion bits of information per second and, when fully loaded with boards, move as many as 60 million packets of data every second.

Repeaters

Repeaters are one of the most simple network interconnection devices. The function of a repeater is to overcome the problems associated with a segment of Ethernet. A repeater also operates in the physical layer of the OSI seven-layer model.

Bridges

A bridge is used to connect two separate LANs or two segments of the same LAN using the same protocols. A bridge is used for three main reasons:

- to overcome physical limitations
- to manage traffic and security
- to allow conversion between technologies.

Bridges operate in the data link layer of the OSI seven-layer model.

Gateways

A gateway is effective in connecting two networks together at any layer at or above the network layer. Gateways are used to overcome problems of incompatibility; however, they can be slow and more expensive than other interconnection devices.

Gateways are effective in small networked environments; for example, in a home office, the home office network can be connected to the Internet, where a nominated machine running the necessary software acts as the gateway.

Wireless devices

Various wireless devices are available to support network interconnection; most are the same as wired devices, such as routers and bridges, while others are more specific, such as wireless adapter cards (Figure 5.10).

Figure 5.10 Wireless adapter: Linksys WMP54G Wireless-G PCI Adapter
http://www.amazon.co.uk/Linksys-Cisco-WMP54G-Wireless-G-Adapter/dp/B00008DOYL

Understand transmission protocols and models

Transmission protocols serve a number of purposes:

- application to application addressing
- reliable data delivery
- segment order maintenance (ensuring that data segments reach the application in the same order that they left)
- flow and congestion control.

There is a range of transmission protocols and models, all providing links at various levels with connection devices.

The OSI model describes how information from a software application on one computer can move through to an application on another computer through a network medium. Any information passing between computers must go through the OSI layers.

The OSI model consists of seven layers, as shown in Table 5.1.

Table 5.1 OSI reference model

Level	Description
7	Applications layer
6	Presentation layer
5	Session layer
4	Transport layer
3	Network layer
2	Data link layer
1	Physical layer

1. **Physical layer** – provides the interface between the medium and the device. The layer transmits bits and defines how the data is transmitted over the network. It also defines what control signals are used and the physical network properties such as cable size and connector, etc.
2. **Data link layer** – provides functional, procedural and error detection and correction facilities between network entities.
3. **Network layer** – provides packing routing facilities across a network.
4. **Transport layer** – an intermediate layer that higher layers use to communicate to the network layer.
5. **Session layer** – the interface between a user and the network, this layer keeps communication flowing.
6. **Presentation layer** – ensures that the same language is being spoken by computers, for example converting text to ASCII and encoding and decoding binary data.
7. **Applications layer** – ensures that the programmes being accessed directly by a user can communicate, e.g. an e-mail programme.

Protocols

There is a range of transmission protocols, including Bluetooth, Wi-Fi, IrDa and cellular radio.

Bluetooth is a short-range wireless communications protocol that enables and facilitates wireless communication via headsets and other mobile devices.

The principle of Bluetooth technology is to use device enquiry and enquiry scan as scanning devices to listen in on known frequencies for other devices that are actively enquiring. As an enquiry is received, the scanning device will send a response showing information needed for the 'enquiring' device to identify the nature of the device that has recognized its signal.

For example, if you wanted to send a wireless picture from one mobile phone to another in the vicinity, you would enable the Bluetooth function on both mobile phone devices, select the picture on your phone and a send option. The phone would begin searching for other Bluetooth-enabled devices in the area. Once the receiving mobile phone has been recognized an acceptance password may be required to accept the connection, then allowing you to transmit the picture (Figure 5.11).

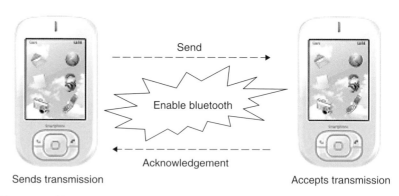

Figure 5.11 Using mobile phone Bluetooth technology (Pictures supplied by Microsoft Clipart)

Wi-Fi

Wi-Fi-enabled devices, which can connect to the Internet when within range of a wireless network that is connected to the Internet, include:

- games console
- MP3 player
- PC
- PDA.

Games consoles have become more dynamic and portable with Wi-Fi-enabled adapters. The article by 'Team Box' discusses the relative benefits of the Xbox 360 wireless networking adapter (Case study 5.1).

Case study 5.1

Xbox 360 Wireless networking adapter review (Xbox 360)

(Cèsar A. Beradini, 29 November 2005). http://hardware.teamxbox.com/reviews/xbox-360/39/Xbox-360-Wireless-Networking-Adapter/p1/

The Xbox 360 Wireless Networking Adapter is one of the most innovate [sic], well designed first-party accessories ever created. Manufacturers of networking gear will have a tough time offering a competitor to this terrific device. Aside from its compact design and Xbox 360 aesthetics, how good [sic] does the adapter perform? We find out.

The Xbox 360 Wireless Networking Adapter is a USB-powered, dual band wireless bridge that has been specifically designed to work in concert with the Xbox 360. Just as gameplay is what really matters in games, all peripherals should be judged by their features and performance. It's after checking those that one realizes how good the Xbox 360 Wireless Networking Adapter really is.

Features

First, its USB interface allowed Microsoft to get away from using an external power supply such as those found in practically all wireless bridges. This is what has allowed the adapter to match the Xbox 360 minimalist design and connect on back of the console without any annoying cables. This also frees up the built-in Ethernet port, and although Microsoft has not yet revealed any intention to use the Xbox 360 as a home router (for which two networking ports would be required), the possibility is at least there and we are sure both Microsoft and Sony have paid attention to that.

The second most important feature of this adapter is its dual band capability. Although 802.11g remains the most popular Wi-Fi flavor, because it offers the same bandwidth found in 802.11a while also being backward compatible with 802.11b, the fact that 802.11g uses the same band that microwave ovens and cordless phones utilize, results in some interference depending on your home construction. If your router supports it, using 802.11a, which works in the 5 GHz range, is always a good alternative.

Finally, the size of this adapter together with its ability to attach on the back of the Xbox 360, make the Xbox 360 Wireless Networking Adapter a must have accessory if you plan to use the Xbox 360 for networking gaming. Speaking of which, it is worth clarifying that the adapter can be used in either infrastructure (for Xbox Live play) or ad hoc mode (for System Link play).

Even if we agreed to focus on features and performance, I must mention again that I'm still impressed by the size of the adapter, not only because of its dimensions, but also of its relative small size for the number of features it packs in. The smallest wireless bridge that I've seen, the Xbox 360 Wireless Networking Adapter is a lot like a notebook wireless bridge in terms of size.

IrDa

The Infrared Data Association defines the physical specifications for communications protocol standards. IrDa interfaces are used in palmtops and mobile phones. Many laptops now tend to use Bluetooth instead of IrDa.

Cellular radio

Cellular radio works on a switching basis. For example, a cellular car phone maintains its connection while the user is mobile by continuously checking for repeater stations and rapidly switching from one station to another to retain the signal. The switching process is so rapid that it does not impede the conversation.

Activity 5.6

There are many transmission standards and protocols, some of which have been referred to in this section. Other protocols include:

- GSM/UMTS
- WAP
- WML
- 802.11
- TCP/IP
- WEP.

For each of the listed standards/protocols identify what they are and provide a brief definition of what they do.

TCP/IP

The TCP/IP model is a four-layer model that consists of the application, transport, Internet and network access layers (Figure 5.12).

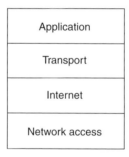

Figure 5.12 TCP/IP model

- **Application layer** – sometimes referred to as the process layer, manages high-level data protocols.
- **Transport layer** – also known as the host-to-host layer. Data is segmented in this layer ready for the next layer. This layer also contains protocols for sending data across the Internet.
- **Internet layer** – segments received from the transport layer are converted into packets and the Internet level delivers the packets that are received.
- **Network access layer** – focuses on all of the issues that an IP packet requires to pass over a physical link from one device to another.

Understand Internet communications

Internet communications embrace a range of technologies, services, hardware and software. Internet communication is also riddled with technical terminology, jargon and abbreviations that could make it inaccessible and unavailable to users at a novice level.

Internet communication

Examples of the terminology associated with Internet communications are HTTP and HTTPS, FTP and SMTP.

Hypertext transfer protocol (HTTP) is a communications protocol that allows information to be transferred or conveyed on intranets or the World Wide Web. HTTPS is a secure, encrypted connection: for example, https://www.payment.co.uk/

File transfer protocol (FTP) is used to transfer data over the Internet or a network from one computer to another. The file transfer process can be used to transmit any type of file (programme, text, graphic, multimedia file, etc.) using a system that bunches the data into packets. When the package of data arrives at its destination, the receiving system/computer checks it to make sure that no errors have been picked up during transmission and then returns a message to confirm receipt of the package and instructions that it is ready to receive another packet (Figure 5.13). The simple mail transfer protocol (SMTP) is the de facto standard for e-mail transmissions across the Internet.

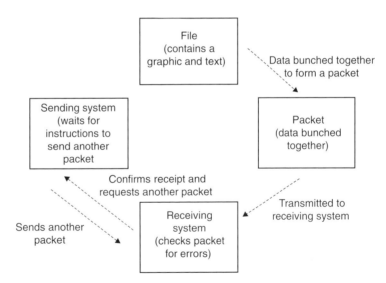

Figure 5.13 File transfer process

Uniform resource locator

The URL is the global address on the World Wide Web for documents and other resources. The address is made up of two parts: the first identifies the protocol to be used and the second identifies the domain name or IP address of the resource.

Example   http://www.train-ed.co.uk

Part 1 (http) – a web page being fetched by the http protocol.
Part 2 (www.train-ed.co.uk) – a web page.

World Wide Web

The World Wide Web or 'web' was created in 1989 as a networked information project and has developed into a series of interlinked, hypertext documents that can be accessed by the Internet.

Blogs and Vlogs

Blogs or web logs are journal-style diary entries that are set up on a website for others to read. Blogs are created chronologically and provide accounts of personal, social, professional, political or other topic areas. Blogs are quite dynamic in that people reading through the entries can sometimes leave their own views and opinions. Vlogs or video logs have the additional feature of video footage that can be used to provide an additional multimedia aspect to any entries.

Wikis

If you want to create and edit web page content freely using any web browser you can use a piece of server software called a wiki. A wiki can be viewed and modified by anybody who has a web browser and access to the Internet, and although they can be open to misuse they can also be quite dynamic in terms of permitting asynchronous communication and group collaboration.

Video conferencing

Video conferencing brings people together over different geographical areas for a meeting, with audio and visual transmission of people, documents and computer-displayed information.

System requirements

System requirements can extend to hardware, software and communication services. Hardware and software can be quite generic for some systems, regardless of whether they are wired or mobile; however, some aspects will be tailored to the type of device and its means of transmission, for example a mobile device such as a PDA would have an antenna, whereas a wired device would not.

Communication services include e-mail, video and the Internet, along with others such as voicemail, conferencing applications, facsimile and collaboration software.

E-mail is the transmission of a message or messages across a communication network. The popularity of e-mail has risen owing to the speed at which messages can be transmitted, and the ability to share data and send multiple copies. Other advantages of e-mail include the cost in relation to using other communication media, the ability to attach a range of multimedia to text documents (pictures, movie clips, hyperlinks, etc.) and auditing: users can store messages (sent and received), track documents by date, time and author, and generate receipts.

Conferencing applications range from text to whiteboarding and more commonly video conferencing. Video conferencing facilities allow users to communicate interactively over a set distance.

Voicemail provides users with the option of setting up a recorded message to capture information that may have been lost if the recipient of the information was not available to take a call.

Facsimile transmits an image through a telephone connection. Faxes can be sent through a conventional fax machine or through a fax modem.

Collaboration software provides the opportunity for groups of users to interact within a secure environment. Collaboration could take the form of an online discussion, verbal communication via a microphone or active participation in the creation or editing of documentation.

Direct communication

Developments in technology have impacted positively on the range of personal communication devices available to consumers. Devices such as laptops, PDAs and mobile phones have enhanced personal communications. In addition, development in technologies has given rise to more direct communications such as chat, video communications, e-mail and web phones (Figure 5.14).

Figure 5.14 Web phones (http://www.apple.com/uk/iphone/)

Direct communication has extended the hours of communication, and the need to communicate with people almost twenty-four hours a day has seen the emergence of virtual communities, and the setting up of newsgroups and chatrooms. In conjunction, the trend for communicating online has seen a growth in the popularity of sites such as MSN Messenger © (Figure 5.15) and Yahoo Messenger © (Figure 5.16).

CHAPTER 5

Figure 5.15 MSN Messenger
http://webmessenger.msn.com/

Figure 5.16 Yahoo Messenger
http://messenger.yahoo.com/

References

The following texts should further enrich your knowledge and understanding of communications technology:

Bartlett, Eugene (2005) *Cable Communications Technology*, McGraw-Hill Professional.

Bensky, Alan (2008) *Wireless Positioning Technologies and Applications*, Artech House.

Elahi, Ata (2005) *Data, Network, and Internet Communications Technology*, Delmar Learning.

Rappaport, Theodore (2002) *Wireless Communications: Principles and Practice*, 2nd edn., Prentice Hall.

Questions and review

1. Provide three examples of communication devices.
2. Explain what the following abbreviations mean:
   - DCE
   - CRC
   - HF
   - USB
   - STP
3. What is meant by the terms bits, bytes and packet structures?
4. How can bandwidth limitations affect digital signalling?
5. What is the difference between 'simplex', 'duplex' and 'half-duplex' communication?
6. Identify the features of three different types of transmission media.
7. Explain what is meant by a network topology and provide three examples of different network topologies.
8. Interconnection devices are components that form the infrastructure of a network. Give three examples of an interconnection device and describe what they do.
9. Provide a detailed overview and a diagram of the OSI 7-layer model.
10. Bluetooth® and Wifi are two examples of transmission protocols. What is meant by a protocol and how do Bluetooth® and Wifi work?
11. TCP/IP is another example of a transmission protocol. Can you explain why TCP/IP is so important?
12. Explain what is meant by the following terminology:
    - HTTP(S)
    - FTP
    - SMTP
    - URL
13. Provide two examples of a communication service.
14. What direct internet communication channels can you identify?

Assessment activities

Grading criteria	Content	Suggested activity
Pass		
P1	Identify and explain types of communication devices.	Design a leaflet that identifies and explains various types of communication devices – both wired and wireless.
P2	Explain the principles of signal theory.	Produce a presentation that explains the principles of signal theory.
P3	Describe communication protocols used and explain why they are important.	Produce a report.
P4	Describe different methods of electronic communication and transmission media used.	Produce an information booklet that could be used to support some seminars at an electronic communication and networking conference. In the first section of the booklet you could describe different methods of electronic communication and transmission media used.
P5	Identify and describe the roles of network components and how they are interconnected.	Another section of the booklet could identify and describe the roles of network components and how they are interconnected.
Merit		
M1	Explain techniques that can be used to reduce errors in transmissions.	Produce a short report that covers M1 and M2. Include sections within the report that cover the techniques used to reduce errors in transmissions for M1.
M2	Explain why particular transmission media are chosen in particular situations.	Another section within the report could explain why particular transmission media are chosen in particular situations.
M3	Explain and implement direct communication between two networked devices.	Demonstrate how direct communication could be implemented between two networked devices and accompany this with a written explanation as to how this was achieved.
Distinction		
D1	Critically compare the OSI seven layer model and the TCP/IP model.	You could produce a table that critically compares the OSI seven layer model with the TCP/IP model. In conjunction to this a written summary could also be included to complement the table.
D2	Choose and justify the choice of a particular access method for a given situation.	Produce a written summary that looks at different access methods, e.g. CSMA/CD, CSMA/CA and token passing. Choose and justify the choice of a particular access method for a given situation – scenarios can be given to cover a range of situations.
D3	Compare and evaluate the effectiveness of the transfer of data over both wireless and wired networks.	In conjunction with P4 and P5 the final section of the conference booklet can compare and evaluate the effectiveness of the transfer of data over both wireless and wired networks.

Courtesy of iStockphoto, vm, Image# 4925279

Keeping systems secure is paramount in today's society. Security can include taking precautions against physical threats by ensuring that rooms and components are locked or password protected. Systems security also includes taking precautions against threats such as viruses, worms and unauthorised access by hackers or other third parties.

With the growth of on-line transactions, the security of data is even more prolific. The transmission of sensitive information such as credit card details over a number of networks has led to advancements in secure payment systems, encryption tools and technologies providing piece of mind and assurances to both consumers and suppliers of ecommerce.

Organizational Systems Security

Computer and data security is crucial in today's society of rapid transmissions and communications across ever-expanding open systems and networks.

The need to store large volumes of data about products, customers, sales, finances, etc., gives rise to concerns about sensitivity and confidentiality: how safe is information?

Users and organizations can safeguard information in a number of ways, using systems ranging from passwords and encryptions to moderators and firewalls. The enforcement of Codes of Practice and compliance with legislation also provide further assurances that frameworks and preventive measures have been set up to address security issues.

This chapter will explore a range of data and system security issues. It will examine the threats to ICT systems and organizations and provide an understanding of how systems and data can be kept secure. Finally, the chapter will look at the issues affecting the use of ICT systems.

The chapter will be structured around the following learning outcomes:

- Know the potential threats to ICT systems and organizations.
- Understand how to keep systems and data secure.
- Understand the organizational issues affecting the use of ICT systems.

Know the potential threats to ICT systems and organizations

Organizations seem to be constantly under threat from a range of internal and external breaches to security. These breaches could derive from malicious destruction to data, hardware or software; and could be a result of natural disasters, viruses, technical failures, human errors, or the theft of data or systems.

Unauthorized access

Unauthorized access to data or entry to systems implies that no permission has been granted from the owner or managers of that data or system. Such access can be external, where somebody is trying to gain access from outside the systems environment, or internal, within the systems environment, possibly an employee within an organization.

Internal and external unauthorized access can pose a threat to ICT systems and organizations. Internally there is the potential of an employee or casual user accessing an area of a system that is not within their remit. This threat can be quite damaging, even if the intention of the user is not malicious. Data could accidentally be erased or edited with dire consequences, especially if there are no passwords limiting users to certain areas. Internal users may also gain access to sensitive or confidential records or information that is possibly reserved for managers or high-end users.

Externally, threats can be sourced by hackers or crackers with the intent of causing malicious damage. Some hackers may not necessarily damage or steal data directly; however, they could infect files and/or the network by jamming resources with a virus, thus disabling certain areas of the system.

Unauthorized access could result in damage to data or jamming of certain resources, such as caused by viruses, or access without data damage, through phishing, identity theft, piggybacking or hacking, where information is viewed and used for other purposes.

Viruses

Viruses are small pieces of software that attach themselves to other programmes and piggyback onto more programmes. Each time a programme runs with a virus attached, the chances are that the virus will multiply and spread throughout a system. Other forms of electronic intrusion and infection include worms, which use computer networks and security holes to replicate, and Trojan horses. Trojan horses claim to do one thing, for example pretend to be a game or a known file, but instead do something else and can cause damage and destruction once opened or released.

Phishing

Phishing is a fraudulent system, whereby someone uses e-mail to send messages to individuals or organizations claiming to be someone that they are not, in order to gain personal data that can then be used for identity theft (Figure 6.1).

Subject: Account Verification
Date: Sun26 Aug 2007 20:43
From: "accountsdepartment"<accounts@.com>

Dear member,
As part of our continuing commitment to protect your account and to reduce the instance of fraud on our website, we are undertaking a period review of our member accounts.
You are requested to visit our site by following the link given below
http://123aanbb-p.com/UpdateInformationConfirm
Please fill in the required information.
This is required for us to continue to offer you a safe and risk free environment to send and receive money online, and maintain your shopping experience.
Thank you
Accounts Management As outlined in our User Agreement, your information about site changes and enhancements. Visit our Privacy Policy and <u>Accounts Agreement</u> if you have any questions.

Figure 6.1 Example of a phishing e-mail

Identify theft

Identity theft is where somebody steals personal information such as your name, address, date of birth and information regarding personal finances for fraudulent purposes such as obtaining a loan or mortgage or entering into some sort of contract (Figure 6.2).

Identity theft

By following a few simple rules you can protect yourself from becoming a victim of identity theft.

Identity theft affects over 100,000 people every year. By obtaining a few of your personal details it is possible for a criminal to open up bank accounts, obtain credit cards, claim benefits and also apply for official documents such as a driving licence, all of which will be traceable to you.

Figure 6.2 Identify theft (http://www.direct.gov.uk/en/RightsAndResponsibilities/DG_10031451)

Activity 6.1

1. What preventive measures can be taken to avoid identity theft?
2. How do you ensure that your identity is kept safe?

Piggybacking

Piggybacking is the method of obtaining free wireless Internet access on the back of somebody else's system. Figure 6.3 provides a case study of this and the implications of partaking in this criminal offence.

Hacking

Hacking into somebody else's system to gain access to data or just for the 'buzz' of knowing that the system is not infallible is almost considered a sport by some individuals. Hacking has become

Man arrested for stealing broadband

ITN - Wednesday, August 22 12:33 pm

A man who was spotted in the street using his laptop to access an unsecured wireless connection has been arrested.

The 39-year-old man was seen sitting on a wall outside a home in Chiswick, west London, by two community support officers.

When questioned he admitted using the owner's unsecured wireless internet connection without permission and was arrested on suspicion of stealing a wireless broadband connection.

The man was bailed to October pending further inquiries.

Dishonestly obtaining free internet access is an offence under the Communications Act 2003 and a potential breach of the Computer Misuse Act.

The move is the latest example of police cracking down on a crime that did not exist several years ago when wireless internet access was relatively rare.

In April, a man was cautioned by police after neighbours saw him using a laptop in a car parked outside a house in Redditch, Worcestershire.

In 2005, a man was fined £500 for piggybacking on someone else's wireless broadband connection in London.

Detective Constable Mark Roberts, of the Metropolitan Police computer crime unit, said anyone who illegally uses a broadband link faces arrest.

He said: "This arrest should act as a warning to anyone who thinks it is acceptable to illegally use other people's broadband connections.

"To do so potentially breaches the Computer Misuse Act and the Communications Act, so computer users need to be aware that this is unlawful and police will investigate any violation we become aware of."

Figure 6.3 Piggybacking case study (http://www.yahoo.co.uk)

glamourized by high-profile cases where hackers have managed to obtain access to military defence systems, Visa and Microsoft.

Damage to or destruction to systems or information

Certain dangers surround the security of systems or information. Damage or destruction to systems or information can come from a number of sources, internal and external.

Natural disasters such as floods and earthquakes cannot be predicted in many cases; however, risk analysis will certainly identify these as dangers and precautions can be taken in terms of more robust, flexible and possibly remote storage and backup procedures. Malicious damage can be from an internal source such as employees working within an organization, or externally from people hacking into the system. Measures such as passwords, encryption, firewall, virus protection and spyware can help to reduce these threats and in many cases ensure that the system remains intruder free.

Other ways in which systems or information can be damaged are through technical failure of hardware, software and communication systems. Technical failure can cause a number of problems for users of the system and also for the organization, as demonstrated in the following case study.

Case Study 6.1

Oyster System Failure

Oyster system failure causes travel misery

Computer fault buggers barriers

By John Oates

Posted in ID, 14 July 2008 09:38 GMT

http://search.theregister.co.uk/?author = John%20Oates

Updated London commuters are suffering more problems than usual this morning, thanks to the weekend failure of the Oyster card readers at tube stations and on buses.

Extra staff have been drafted in this morning to sort out problems with cards which were used between 5.30am and 10.30am on Saturday. Some cards were apparently wiped, meaning some customers were left with cards that didn't work and/or a fine.

Although the problems were supposedly fixed by 10.30am Saturday morning, we've had reports of problems on buses up until late Saturday evening.

Certain Freedom Passes and Young Persons Oyster cards might also need to be exchanged for a new card. Transport for London believes as many as 40,000 cards might have problems.

TfL said the computer problem was fixed at 9.30am on Saturday but some retailers did not get ticket services back until Sunday. People who were charged a maximum fare on Saturday morning will get an automatic refund on Tuesday.

The problem comes just weeks after Dutch researchers found a way to clone the Mifare chip which the card is based on. Chip company NXP is taking legal action to silence the university researchers who revealed the problem.

Questions

1. What happened to the Oyster system?
2. What caused the suspected failure?
3. What was the impact on the customers?

Human error can cause a range of system problems, some of which can be critical depending on the environment. Mistakes can be made by users at the input, process or output stage. Theft is also an issue, in terms of both data theft and physical system theft, which can be combated by more electronic and physical security.

Information security

Keeping information secure is difficult as hackers are always inventing new ways of gaining access to systems and releasing viruses, worms and trojans. Maintaining confidentiality of information is a big issue for organizations, especially with the large volumes of data that are stored about customers, clients, patients and students, which contain sensitive and personal information. Although there is legislation in place to protect and offer a formal protective measure, this should be in addition to more proactive and hands-on techniques. Other considerations in terms of keeping information secure include the integrity and completeness of data. Can you trust the source of the data? Is it free from bias, etc., and is it available when needed?

Threats related to e-commerce

With the growth of online trading and the billions of pounds worth of transactions passing between consumers and businesses every day, threats to websites, concerns over data access issues and denial of service attacks force e-commerce traders to be more prudent. The threats relating to e-commerce include:

- **website defacement** – a hacker may gain access and alter the HTML code, change the graphics or maliciously cause damage by infecting the system with a virus or installing spyware
- **control of access to data via third party suppliers** – measures need to be taken if access is required to a third party site; for example, for a payment system or an external stock system, in terms of sensitive data being passed between two or more sites
- **denial of service attacks** – multiple connections are generated to overload the server or the network, thus disabling the website.

Counterfeit goods

Another threat to ICT systems, electronic communication and organizations comes in the form of counterfeit goods that can increase security risks, especially with regard to software, DVDs, music and games. Counterfeit goods may contain viruses that can cause serious damage if downloaded or installed onto a system, especially if it is networked. From an organizational point of view, the purchase of counterfeit goods can also cost the company money and force it to increase prices to counteract any losses due to piracy, etc., thus ultimately impacting on the consumer.

Organizational impact

For any organization that uses ICT as a storage, processing, output and communication device the risks and threats to the system are far reaching. There are various threats, both internal and external, as previously explored, that can be addressed through a range of preventive and physical measures.

Failure to act can result in loss of service if the site is down, possible data loss and loss of business or income. If a site is infiltrated and damaged maliciously, the corporate image could be affected, especially if data is lost or if consumers cannot access information. Therefore, it is essential that security is at the forefront of any planning, budgeting and operational decision making at all levels.

Understand how to keep systems and data secure

The need to protect systems and IT resources is paramount within any organization and because of this many organizations invest large amounts of money in the development and implementation of a security policy.

Physical security

A number of physical security measures can be taken to ensure that systems and data are kept secure. These range from using physical locks both on hardware components, chaining these to desks, etc., and on cupboards and doors to secure the environment in which the systems are physically stored. Visitor passes can also be used to monitor access to certain areas within a building and ensure only authorized access. In conjunction with this, visitors and employees could use a signing-in procedure at a front desk to monitor traffic within an environment. Other forms of security measures include guards and cable shielding, to stop electrical interference 'cross-talk'.

Biometrics

Biometrics is a way of identifying somebody through physiological or behavioural characteristics such as facial expressions, fingerprinting, voice recognition or body scanning.

One organization that has embraced biometrics is the Home Office, in terms of issuing new 'biometric' passports. The passports contain a chip that holds a range of personal data (Figure 6.4).

Figure 6.4   Biometric passports (http://www.passport.gov.uk/general_biometrics.asp).

CHAPTER 6

Activity 6.2

Biometrics is prevalent in a range of products and technologies.

1. Carry out research to identify two different ways in which biometrics have been used.

Biometric Passports

What information is stored on the chip?

The chip stores the passport holder's photo and the personal details printed on page 31 of the passport.

Can I see what is on the chip?
IPS Regional Offices in Belfast, Glasgow, Durham, Liverpool, Peterborough, London and Newport are now equipped with Biometric Passport Readers. The Reader enables holders of a British biometric passport to view their personal information stored within the chip embedded in the passport. The information viewed is the same as the information displayed on the bio data page of the passport.

This is a free self service facility within the public area of each IPS office.

Information about Regional Office opening hours.

2. Provide details on the product, type of biometric technology used and any other information that may be appropriate.
3. Below is an example of a biometric product, a biometric door handle that uses fingerprinting to activate the lock controls of the handle. Think about another everyday item that could incorporate biometric technology and identify a way in which biometrics could be incorporated into the product.

Voice recognition is another way of ensuring that data can be kept secure. A user speaks into a microphone, and the voice pattern is analysed, stored and used to lock or secure a certain file, folder or system. Only the same voice recognition pattern can activate or gain access to the required area.

Software and network security

Security of data is paramount for an organization, and the methods used to counter unauthorized intrusions and protect systems include encryption techniques, public and private key, callback, handshaking, diskless networks, backups, audit logs, firewalls, use of virus-checking software and virtual private networks, intruder systems and passwords. In addition, setting access levels and updating software on a regular basis are ways of ensuring that systems and data are kept secure.

Keeping data secure can be difficult because of the environment in which users work, and levels of user and access requirements to the data.

With the movement towards a totally networked environment promoting a culture of sharing, the issue of data security is even more important and should be addressed at a number of levels (Figure 6.5). Security

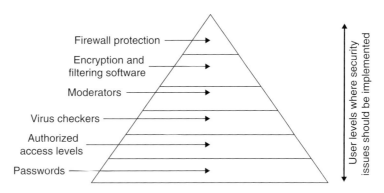

Figure 6.5 Levels of security

measures need to be integrated at each user level within an organization. The indication of security measure does not confine it to a certain level, but reflects on an organizational scale that should be implemented and the scale of implementation. In addition to these proposed security measures is the issue of physical security, ensuring that hardware and software are kept physically secure, under lock and key.

The actual protection of data can be resolved easily by introducing good practice measures such as backing up all data to a secondary storage device, limiting file access, and imposing restrictions to read only, execute only or read/write. However, data protection is also covered more widely under certain Acts such as the Data Protection Act 1984.

Encryption and filtering software

Encryption software scrambles message transmissions. When a message is encrypted a secret numerical code is applied, the 'encryption key', and the message can be transmitted or stored in indecipherable characters. The message can only be read after it has been reconstructed through the use of a 'matching key'.

Encryption can be categorized in terms of being 'private' key or 'public' key. Private key encryption ensures that both the sender and recipient of data have a key that is used for both encryption and decryption. With public key encryption two different but related keys are used to encrypt and decrypt data. Data encrypted with the first key can only be decrypted by the second key.

One of the most common examples of public key encryption is the secure sockets layer (SSL) protocol, which is commonly used in web programmes and transactions.

Handshaking

Handshaking is a process for establishing the rules and protocols of communication between devices. For example, if a computer needs to

CHAPTER 6

communicate with another device such as a printer it has to handshake initially to establish connection.

Diskless network

A diskless network can be more secure than networks with a disk facility because the network is not compromised by users. Users cannot load data or information onto the network directly, therefore removing the risk of viruses and other malicious programmes being transferred onto the system.

Backups

The consequences of an organization losing data can be very severe and the disruption to processing activities costly and time-consuming. Attempts to retrieve lost data (if retrievable) can prove to be a fruitless exercise, putting strain on both manual and financial resources.

Backup options available to an organization include:

- simple backup
- stack backup
- advanced stack backup
- incremental backup
- grandfather, father, son backup.

Simple backup is the elementary backup type. Each time an archive is created the oldest version of the backup file is replaced with the newly created one.

Stack backup consists of the last created backup and previous versions, the previous versions being organized into a stack format.

The advanced backup procedure differs in that it does not permit unchanged or unedited files in the old backup version copies to be stacked.

An incremental backup provides a fast method of backing up data much more quickly than a full backup. During an incremental backup only the files that have changed since the last full or incremental backup are included, and therefore the time it takes to conduct the backup may be a fraction of the time it takes to perform a full backup.

The grandfather, father, son technique is probably the most common backup method, which uses a rotating set of backup disks or tapes so that three different versions of the same data are held at any one time. An example of this method is shown in Table 6.1.

Table 6.1 Grandfather, father, son backup method

Stock order data					
Wednesday		Thursday		Friday	
Disk 1	Grandfather	Disk 2	Grandfather	Disk 3	Grandfather
Disk 2	Father	Disk 3	Father	Disk 1	Father
Disk 3	Son	Disk 1	Son	Disk 2	Son

Firewall protection

The primary aim of a firewall is to guard against unauthorized access to an internal network. In effect, a firewall is a gateway with a lock, and the gateway only opens for information packets that pass one or more security inspections.

There are three basic types of firewall:

- **Application gateways** – the first gateways, sometimes referred to as proxy gateways. These are made up of hosts that run special software to act as a proxy server. Clients behind the firewall must know how to use the proxy, and be configured to do so in order to use Internet services. This software runs at the 'application layer' of the ISO/OSI reference model, hence the name. Traditionally, application gateways have been the most secure, because they do not allow anything to pass by default, but need to have the programmes written and turned on in order to begin passing traffic.
- **Packet filtering** – a technique whereby routers have 'access control lists' turned on. By default, a router will pass all traffic sent it, and will do so without any sort of restrictions. Access control is performed at a lower ISO/OSI layer (typically, the transport or session layer). Because packet filtering is done with routers, it is often much faster than application gateways.
- **Hybrid system** – a mixture between application gateways and packet filtering. In some of these systems, new connections must be authenticated and approved at the application layer. Once this has been done, the remainder of the connection is passed down to the session layer, where packet filters watch the connection to ensure that only packets that are part of an ongoing (already authenticated and approved) conversation are being passed.

Moderators

Moderators have the responsibility of controlling, filtering and restricting information that is shared across a network.

Virus checkers

These programmes are designed to search for viruses, notify users of their existence and remove them from infected files or disks.

Authorized access levels and passwords

On a networked system various privilege levels can be set up to restrict users' access to shared resources such as files, folders, and printers and other peripheral devices. A password system can also be implemented to divide levels of entry according to job role and information requirements. For example, a finance assistant may need access to personnel data when generating the monthly payroll. Data about employees, however, may be password protected by personnel in the human resources department, so special permissions may be required to gain entry to this data.

Audit control software will allow an organization to monitor and record what they have on their network at a point in time and provide them with an opportunity to check that what they have on their system has been authorized and is legal.

Over time, a number of factors could impact on how much software an organization acquires without its knowledge, such as:

- illegal copying of software by employees
- downloading of software by employees
- installation of software by employees
- exceeded licence use of software.

These interventions by employees may occur with little or no consideration to the organization and its responsibility to ensure that software is not misused or abused.

User rights and file permissions

Within certain IT systems, users are given permission to access some areas of a folder, an application or a document, but are restricted from others. By allowing users certain rights within a given system, security of data can be reassured and the span of control can be limited. For example, in the case of an IT system in a doctor's surgery, the administration staff may have access to appointments and scheduling, the nurses may have access to patient information, and the GP could have full access and rights to print out prescriptions and authorize medication.

Certain permissions may also be set up to allow particular users partial access to a file; so, for example, information can be read (read only), but not written to, or users may be able to run a programme (execute only), but not view it. Users with full read/write permissions would be able to view, update, amend and delete accordingly.

Understand the organizational issues affecting the use of ICT systems

A range of organizational issues can affect the use of ICT systems. Some of these issues can be addressed through the enforcement of Codes of Practice and Codes of Conduct. Having risk assessment and disaster recovery plans can also address some of the ICT concerns. Compliance with legislation and professional bodies and regulatory authorities can also have an impact on the way an organization operates and manages its ICT provision.

Security policies and guidelines

The contents of a corporate information systems policy will vary between organizations; however, there are some common features that will form the basis of a policy. Once potential threats and risks have been identified, policies can be put in place to address them. A corporate ICT security policy can be drawn up to address such threats, and will typically include:

- **access controls** – identifying and authenticating users within the system, setting up passwords, building in detection tools, encrypting sensitive data
- **administrative controls** – setting up procedures with personnel in case of a beach; disciplinary actions, defining standards and screening of personnel at the time of hiring
- **operations controls** – backup procedures and controlling access through smart cards, log-in and log-out procedures, and other control tools
- **personnel controls** – creating a general awareness among employees, providing training and education
- **physical controls** – securing and locking hardware, having another backup facility offsite, etc.

Corporate information systems security policy

Contingency plans can be used to combat potential and actual threats to a system. The majority of organizations will have an adopted security policy of which employees would be aware, the plan and policy being open to continuous review and updating.

The structure of a contingency plan is unique to the organization and its requirements. Some organizations are more at risk than others, depending on:

- size
- location
- proximity to known natural disasters and threats: flood areas, etc.
- core business activity.

Disaster recovery plans

A strategy based on recovery recognizes that no system is infallible. As a result, numerous companies have emerged providing 'disaster recovery' services if no internal organization contingency plan has been drawn up. These companies maintain copies of important data and files on behalf of an organization.

A disaster recovery plan may include a provision for backing up facilities in the event of a disaster. These provisions may include:

- subscribing to a disaster recovery service
- an arrangement with another company that runs a compatible computer system
- a secondary backup site that is distanced geographically from the original
- the use of multiple servers.

Some large organizations may have a 'backup site' so that data processing can be switched to a secondary site immediately in the case of an emergency. Smaller organizations may employ other measures such as RAID or data warehousing facilities.

CHAPTER 6

Activity 6.3

1. What security measures can be enforced within an organization?
2. What is meant by the term 'risk analysis'?
3. Why are some organizations more at risk than others in terms of potential threats to their systems?
4. What measures can a large and a small company take to protect their data?

Auditing allows an organization to take stock of what they have at a point in time. It provides them with an opportunity to check that what they have on their system has been authorized and is legal, and to monitor the activities of employees, for example illegal copying or downloading of software.

In conjunction with audits on software, audits can be carried out by identifying and correcting data. A data quality audit is a structured survey of the accuracy and level of completeness of data within a system. This type of audit can be carried out using the following methods:

- surveying end-users to gather their perceptions on data quality
- surveying samples gathered from data files
- surveying entire data files.

Unless regular data quality audits are undertaken, organizations have no way of knowing the extent to which their information systems contain inaccurate, incomplete or ambiguous information.

Audit trails are carried out for a number of reasons, including checking users to see what information has been changed or updated, when and by whom. Audit trails are also written into applications and recorded with each transaction.

Employment contracts and security

Employees can be quite a security liability for an organization; therefore, rigorous checks and procedures need to be carried out in the selection or hiring stages and monitoring should continue through employment.

Employees should be aware of various security policies and procedures in terms of Internet and e-mail usage and guidelines on downloading and installing software. Some of these policies can be contained within a Code of Practice or Code of Conduct that sets out conditions for ICT use.

Training is an essential part of communicating policies and procedures and ensuring that employees are skilled and knowledgeable in their job roles. The benefits of training include:

- a skilled and up-to-date workforce
- a motivated workforce
- greater efficiency, more productivity and a more competitive organization

- a more flexible workforce that can adapt to changes in job tasks and roles
- improved and enhanced skills and knowledge that can be applied to individual job roles.

At an operational level training would be more practically based:

- application software training
- inventory and stock control
- accounts and payroll
- health and safety.

The majority of training at this level may involve a number of employees being trained at the same time, possibly inhouse, to cut costs.

The vast majority of training in an organization will focus around software. Applications training is probably the most sought-after training requirement because of the generic skills required by users across a range of functional areas. A number of recognized training methods can be used to instruct and inform users, especially for applications training. These include:

- instructor-led training
- hands-on training
- online training
- self-teaching
- user training manual.

Instructor-led training can be carried out inhouse or outsourced to a third party. This type of training is usually carried out by a professional who specializes in a particular area such as applications software, programming, Internet, web applications or hardware. The instructor will lead the event and help users through the use of demonstrations, practical activities and examples.

Hands-on training is mainly based around observation, watching somebody doing the task, and then carrying out the task using the skills and knowledge of the person who was being observed. This type of training can also be referred to as 'on the job'.

Online training is extremely popular because it can be carried out, in some cases, twenty-four hours a day. Users can log onto a particular site, download material, and work through examples and activities at their own pace.

Self-teaching can also be linked in with user training manuals, where individuals learn or update skills by reading material and applying what they have learnt to a range of exercises or practical activities. This method of training is especially popular for ICT professional qualification status in terms of becoming Microsoft or Cisco accredited.

The way in which training can be delivered will depend on an organization's training strategy. Some organizations develop their own training strategy to satisfy a specific time-frame, e.g. over a period of six months or a year. This strategy may include:

- **training type** – hardware, software, specialist, e.g. payroll or health and safety, management training, etc.
- **personnel to be trained** – all new employees, functional departments, supervisors, etc.
- **delivery format** – instructional, observational, self-teaching, online
- **location** – inhouse, residential, day release.

Another training strategy option would be to outsource to other companies, especially if the level of training required is quite specialist.

It is beneficial to both employees and employers for skills to be continually updated and refreshed. For an employee, skills updating may provide the opportunity to move into a new job role, or make them more marketable for another job. For an employer, an up-to-date trained workforce could reduce costs and inefficiencies, and make them more competitive.

Code of Conduct

Codes of Practice and Codes of Conduct provide guidelines about the standards and quality of work to be undertaken by employees within an organization.

The code is intended to ensure that a high level of professionalism is maintained when working within the ICT industry. A Code of Practice contains a number of elements, which may include:

- responsibilities for use of company hardware and software
- responsibilities for the use of data
- responsibilities for the correct use of time
- responsibilities for the use and control of the Internet or company intranet system
- authorization in terms of security, passwords and access rights, etc., depending on the level of the employee
- the implementation of legislation such as the Data Protection Act.

A Code of Practice helps to clarify the expectations and behaviour of employees and employers on a professional level at work. This code is not part of the legislative procedure; but it is still enforceable within organizations. Within an ICT environment the need for this code is important so that issues of abusing resources in worktime are addressed and boundaries that are acceptable to both parties are agreed.

All employees should understand and distinguish between right and wrong, correct and incorrect procedures and behaviour. To ensure that this distinction is upheld and to maintain a consistent level of professionalism within the workplace, organizations release a Code of Conduct.

The Code of Conduct should outline what is acceptable within an employee's work environment in terms of how they:

- carry out their job role
- interact with other employees
- represent themselves professionally
- communicate with third parties.

These areas establish and define an individual's duties to the profession, employer and clients, and their own levels of competency, integrity and honesty.

Many organizations set out Codes of Conduct to ensure that standards are being maintained and that employees are aware of their responsibilities and how they should behave at work. Codes of conduct are not exclusive to employees of organizations: a range of other codes exists for members of organizations, such as the British Computer Society (BCS) (Case study 6.2).

Case study 6.2

The British Computer Society Code of Practice

British Computer Society
Code of Practice

Introduction

The British Computer Society sets the professional standards of competence, conduct and ethical practice for computing in the United Kingdom. The Society was incorporated by Royal Charter in July 1984.

This Code of Practice is directed to all professional members of The British Computer Society and was issued in September 1999. It is a generic Code of Practice, consisting essentially of a series of statements which prescribe minimum standards of practice, to be observed by all members.

The Code is concerned with professional responsibility. All members have responsibilities: to clients, to users, to the State and society at large. Those members who are employees also have responsibilities to their employers and employers' customers and, often, to a Trade Union. In the event of apparent clash in responsibilities, obligations or prescribed practice the Society's Registrar should be consulted at the earliest opportunity.

The Code is intended to be observed in the spirit and not merely the word. The BCS membership covers all occupations relevant to the use of computers and it is not possible to define the Code in terms directly relevant to each individual member.

The General Code of Practice

Purpose

A Code of Practice is a set of principles established to ensure that Information Systems (IS) Practitioners maintain recognized standards of competence in all aspects of IS practice. The Code of Conduct establishes the principles of behaviour in IS practice.

This General Code of Practice has been introduced to:

- Provide guidance to all Information Systems Practitioners regardless of their specialism.
- Allow for the provision of sector-specific Codes of Practice governing practice in specialist areas.

Employers, clients, members of the public or other professions should expect the same high standards of practice in Information Systems as are expected from members of other recognized professions. Professional Institutions set standards of technical capability which professional Practitioners must satisfy, and they prescribe standards of professional practice to which their members must conform or be held accountable for any lapse.

Scope

This General Code of Practice applies to all Professional Members of the British Computer Society (BCS) and to such specialist communities of Practitioners who choose to adopt a sector-specific Code.

Responsibilities of Information Systems Practitioners

Information Systems Practitioners shall seek to upgrade their professional knowledge and skill and shall maintain awareness of technological developments, procedures and standards which are relevant to their field, and shall encourage their subordinates to do likewise.

The Practitioner has a duty of professional care to abide by those BCS sector-specific Codes of Practice which have any bearing on the Practitioner's current role.

Should the Practitioner discover any difficulty in applying a Code of Practice, he or she shall immediately bring it to the attention of the BCS Registrar.

Information Systems Practitioners who are BCS members adopting the General Code of Practice are also bound by the requirements of the BCS Code of Conduct. Others adopting this Code of Practice are also advised to adhere to the Code of Conduct.

Authorization

This General Code of Practice is approved by Council with the authority of Qualifications and Standards Board.

Sector-specific Codes to be adopted must be:

- Authorized by an appropriate specialist community, then
- Approved by Qualifications and Standards Board.

Development and maintenance of Codes of Practice

All BCS-initiated Codes of Practice are maintained through Qualifications and Standards Board. In the event that the BCS adopts Codes of Practice originating from other recognized bodies, these will be maintained by the originating bodies.

Activity 6.4

1. **Why should organizations have a Code of Practice?**
2. **What would a typical Code of Practice outline?**

Laws

Legislation can affect different users in different ways. Organizations have to ensure that they operate within certain legislative boundaries, which include informing employees and third parties about how they intend to safeguard systems and any information collected, processed, copied, stored and output on these systems. Personal users need to be aware of the required legislation, what their responsibilities are and the penalties that can be imposed if any laws or regulations are breached.

The types of legislation that an organization needs to consider affect everyday operations in terms of:

- collecting, processing and storing data
- using software
- protecting their employees and ensuring that working conditions are of an acceptable standard.

Computer Misuse Act 1990

The Computer Misuse Act was enacted to address the increased threat of hackers trying to gain unauthorized access to a computer system. Prior to this Act there was minimal protection and prosecution was difficult because theft of data by hacking was not considered as deprivation to the owner. Offences and penalties under this Act are listed below.

Offences

- **Unauthorized access** – an attempt by a hacker to gain unauthorized access to a computer system.
- **Unauthorized access with the intention of committing another offence** – on gaining access, a hacker will then continue with the intent of committing a further crime.
- **Unauthorized modification of data or programmes** – introducing viruses to a computer system is a criminal offence. Guilt is assessed by the level of intent to cause disruption, or to impair the processes of a computer system.

Penalties

- **Unauthorized access** – imprisonment for up to six months and/or a fine of up to £2000.
- **Unauthorized access with the intention of committing another offence** – imprisonment for up to five years and/or an unlimited fine.
- **Unauthorized modification** – imprisonment for up to five years and/or an unlimited fine.

Copyright

Copyright is awarded to a product or brand following its completion. No further action is required in order to activate it. Copyright is transferable; the originator/author grants it and it can exist for up to fifty years following the death of the originator/author.

CHAPTER 6

A number of copyright issues exists with regard to software:

- **Software piracy** – the copying of software to be used on more machines than individual licences have been paid for.
- **Ownership** – if a bespoke piece of software has been developed for an organization the copyright remains with the developer unless conditions have been written into a contract.
- **Transference** – can an employee who has developed a piece of software for their organization take the ownership and copyright with them to another? Unless this is addressed in the employee's contract the organization will have no right to any software developed.

The Federation Against Software Theft (FAST) was set up in 1984 by the software industry with the aim of preventing software piracy. Anybody caught breaching the copyright law will be prosecuted under this federation.

Copyright, Designs and Patent Act

The Copyright, Designs and Patent Act provides protection to software developers and organizations against unauthorized copying of their software, designs, printed material and any other product. Under copyright legislation an organization or a developer can ensure that its intellectual property rights (IPR) have been safeguarded against third parties who wish to exploit and make gains from the originator's research and development.

Data Protection Act

The Data Protection Act applies to the processing of data and information by any source, either electronic or paper based. The Act places obligations on people who collect, process and store personal records and data about consumers or customers. The Act is based on a set of principles that binds a user or an organization into following a set of procedures offering assurances that data is kept secure.

The main principles are:

1. Personal data shall be processed fairly and lawfully and, in particular, shall not be processed unless:
   - at least one of the conditions in Schedule 2 of the 1998 Act is met, *and*
   - In the case of sensitive personal data, at least one of the conditions in Schedule 3 of the 1998 Act is also met.
2. Personal data shall be obtained only for one or more specified and lawful purposes, and shall not be further processed in any manner incompatible with that purpose or those purposes.
3. Personal data shall be adequate, relevant, and not excessive in relation to the purpose or purposes for which they are processed.
4. Personal data shall be accurate and, where necessary, kept up to date.
5. Personal data processed for any purpose or purposes shall not be kept for longer than is necessary for that purpose or those purposes.
6. Personal data shall be processed in accordance with the rights of data subjects under this Act.

7. Appropriate technical and organizational measures shall be taken against unauthorized or unlawful processing of personal data and against accidental loss or destruction of, or damage to, personal data.

8. Personal data shall not be transferred to a country or territory outside the EEA (European Economic Area) unless that country or territory ensures an adequate level of protection for the rights and freedoms of data subjects in relation to the processing of personal data.

The Act gives rights to individuals in respect of personal data held about them by data controllers. These include the rights:

- to make subject access requests about the nature of the information and to discover to whom it has been disclosed
- to prevent processing likely to cause damage or distress
- to prevent processing for the purposes of direct marketing
- to be informed about the mechanics of any automated decision-taking process that will significantly affect them
- not to have significant decisions that affect them made solely by an automated decision-taking process
- to take action for compensation if they suffer damage by any contravention of the Act by the data controller
- to take action to rectify, block, erase or destroy inaccurate data
- to request the Commissioner to make an assessment as to whether any provision of the act has been contravened.

The Act provides wide exemptions for journalistic, artistic or literary purposes that would otherwise be in breach of the law.

The role of the Data Protection Commissioner

The Commissioner is an independent supervisory authority and has an international role as well as a national one. Primarily, the Commissioner is responsible for ensuring that the Data Protection legislation is enforced.

In the UK, the Commission has a range of duties, including:

- promotion of good information handling
- encouraging Codes of Practice for data controllers.

To carry out these duties the Commissioner maintains a public register of data controllers. Each register entry contains details about the controller, such as their name and address and a description of the processing of the personal data to be carried out.

Registering entries

All users, with a few exceptions, have to register an entry or entries giving their name, address and broad descriptions of:

- those about whom personal data is held
- the items of data held
- the purposes for which the data is used

Personal data – information about living, identifiable individuals. Personal data does not have to be particularly sensitive information and can be as little as name and address.

Data users – those who control the contents, and use of, a collection of personal data. They can be any type of company or organization, large or small, within the public or private sector. A data user can also be a sole trader, a partnership or an individual. A data user need not necessarily own a computer.

Data subjects – the individuals to whom the personal data relates.

Automatically processed – processed by computer or other technology such as documents image-processing systems.

Open-source software has human-readable source code that is available within the public domain to users who can then use, change or improve it.

Freeware software is similar to open-source software in that it is available for free. However, unlike open-source software it is usually only available in a binary format so that it cannot be changed in any way shape or form by users.

Shareware software is similar to freeware; however, the trial version is usually free but if you want access to the full programme a registration fee is required. A good example of this is 'Doom' or anti-virus software.

Commercial software is designed to be bought by users; for example, utilities, games, education, finance or office-type packages.

- the sources from which the information may be disclosed, i.e. shown or passed to
- any overseas countries or territories to which the data may be transferred.

Copyrights

Several different copyright licensing methods provide users with open access to certain pieces of software that is royalty free or available to use under certain conditions, for example:

- open source
- freeware
- shareware
- other commercial software.

Activity 6.5

1. What are the benefits and limitations of freeware and open-source software?
2. Provide two examples of shareware software.

Ethical decision making

Ethical decision making can force organizations into a dilemma over 'freedom of information' versus 'personal privacy'. In addition, organizations that pool information on individuals from electoral rolls and other organizations selling data to compile a commercial register may be considered unethical and in breach of personal rights.

Even the use of ICT can cause a number of ethical and moral issues. The use of technology to capture, process, store, manipulate and output information has in many cases improved operations and efficiency levels in organizations, but there are many issues associated with the way in which ICT has been introduced and managed.

Ethical and moral issues arise in two areas. First, such issues concern the way in which data and information are stored and used on ICT systems, in terms of:

- what data is being stored
- whether the data storage has been authorized by the person on whom the data is based
- who has the right to see and use the data.

Secondly, ethical and moral issues arise in terms of the use of general ICT systems by employees and the measures taken by employers to monitor such use. Questions arise in terms of:

- Should employees use ICT resources for their own personal gain during work?
- Do employers have the right to monitor their employees?

- To what extent should monitoring take place?
- Do employees need to be informed of monitoring activities?

Professional bodies

Several professional bodies uphold standards and support or contribute in some way to organizational system security; for example, the Business Software Alliance (BSA), FAST, BCS and the Association of Computing Machinery (ACM).

The BSA was set up in 1988 as a trade group representing some of the world's largest software houses and manufacturers. The primary objective of the BSA is to stop the copyright infringement of software, which causes billions of pounds worth of losses to the developers and manufacturers.

FAST (http://www.fastis.org.uk) was set up in 1984 by the software industry with the aim of preventing software piracy. Anybody caught breaching the copyright law will be prosecuted under this federation.

The BCS (http://www.bcs.org) is a professional body for those working in IT. As a professional body it offers resources and support to its members and requires its members to abide by a Code of Conduct that provides professional practice guidelines.

The ACM (http://www.acm.org) provides a support and resource body for advancing computing as a science and profession.

Activity 6.6

1. Look at the BCS and ACM websites and identify common areas of support offered to their members.
2. Identify the fees associated with membership for each.
3. Why do you think that a fee is required? Should membership be free?

CHAPTER 6

References

The following texts should further enrich your knowledge and understanding of organizational systems security:

Barman, Scott (2001) *Writing Information Security Policies*, Sams Publishing.

Lopez, Javier (2008) *Securing Information and Communications Systems: Principles, Technologies, and Applications (Information Security & Privacy)*, Artech Publishing.

Williams, Paul (2008) *Security Studies: An Introduction*, Routledge.

Questions and review

1. There are a number of potential threats to ICT systems. One of these threats includes unauthorised access. Can you identify three threats that can be caused by unauthorised access to a system?
2. An IT system can also be at risk from a range of physical and natural threats. Can you identify a range of these and also describe what measures can be taken to address or overcome these threats?
3. What factors do you need to consider in terms of information security?
4. What threats might you encounter if you had an e-commerce provision?
5. Counterfeit goods are popular because they are cheaper and possibly more widely available than the original product. What products might be at risk under this category?
6. What impact can threats to ICT systems have on an organisation? What measures could an organisation take to pre-empt these threats?
7. What physical measures can be taken to protect ICT systems?
8. What is meant by the term 'biometrics'? Provide an example of biometrics in use in today's society
9. There are a number of ways to keep networks and the software on them secure. Can you identify three ways in which this can be done?
10. Why is it a good idea for an organisation to have a 'disaster recovery policy' and what might this policy include?
11. Do you think that employees should abide by a code of conduct? Justify your answer.
12. Explain what is meant by the following copyright terms:
    - Open source
    - Freeware
    - Shareware
13. What ethical decisions might an organisation have to make regarding the use of IT systems?
14. What do the following professional bodies do/represent?
    - BSA
    - FAST
    - BCS
    - ACM

Assessment activities

Grading criteria	Content	Suggested activity
Pass		
P1	Describe the various types of threats to organisations, systems and data.	You could produce a report that covers P1, P2, P3, P4, M1, M2 and D1. The report should be quite substantial at between 3,000–3,500 words and should address security threats, their impact and countermeasures.
		For P1 produce a section within the report that describes the various threats to organisations.
P2	Describe the potential impact of four different threats.	The report should also include the potential impact of four different threats.
P3	Describe the countermeasures available to an organisation that will reduce the risk of damage to information.	Countermeasures could be covered within a combined section that would reduce the risk of damage to information.
P4	Describe the countermeasures available to an organisation that will reduce the risk of damage to physical systems.	Countermeasures could be covered within a combined section that would reduce the risk of damage to physical systems.
P5	Describe different methods of recovering from a disaster.	Produce a presentation aimed at informing organisations about policies and procedures required to ensure ICT systems security. The first part of the presentation could look at different methods of recovering from a disaster.
P6	Describe the tools and policies an organisation can adopt in managing organisational issues in relation to ICT security.	The presentation could also describe various tools and policies that an organisation can adopt to manage organisational issues in relation to ICT security.
P7	Describe how staff contracts and code of conduct can assist the task of ensuring secure systems.	Write a leaflet detailing how staff contracts and code of conduct can assist the task of ensuring systems.
Merit		
M1	Explain possible security issues which exist within a given system.	Within the report there should be a section that addresses security issues. You should outline and explain the possible security issues that may exist within a given system.
M2	Explain the operation and effect of two different threats involving gaining access to information without damage to data.	Extend the section on security issues to explain the operation and effect of two different threats involving gaining access to information without damage to data.
M3	Explain the operation and use of an encryption technique in ensuring security of transmitted information.	The presentation could provide an explanation of the operation and use of an encryption technique to ensure the security of transmitted information.
Distinction		
D1	Describe the possible security issues which exist within a given system identifying the likelihood of each and propose acceptable steps to counter the issues.	The final section of the report should describe possible security issues that exist within a given system and identify the likelihood of each issue proposing acceptable steps to counter the issues.
D2	Justify the security policies used in an organisation.	The presentation could conclude with a possible case study example that justifies the security policies used in an organisation.

CHAPTER 6

Courtesy of iStockphoto, kr7ysztof, Image# 2675000

Software can be designed and developed in a number of ways using a range of languages, tools and applications.

Software design can be influenced by a number of factors such as, the type of software to be designed, for example is it commercially based to suit a specific target audience, or is it more generic for a range of end users. Cost is also important because a limited budget may reflect on the complexity of the design. Time should also be factored into the development process, the more time that is allocated to a software project the more time can be allocated to researching, drafting and testing the product.

Chapter 7

Principles of Software Design and Development

This chapter will provide you with an overview of software design and development processes and principles.

Software solutions can be quite simplistic, based on the use of applications software and utilities that can be used to support a range of business functions. Some software solutions, however, may need to be more tailored, and require the planning and build of a new programme to support a given business function. Whatever method is selected, the mechanics of developing, implementing, testing and documenting the software should follow a similar cycle and this will be explored further within the chapter.

The chapter will be structured around the following learning outcomes:

- Know the nature and features of programming languages.
- Be able to use software design and development tools.
- Be able to design and create a programme.
- Be able to document, test, debug and review a programmed solution.

Know the nature and features of programming languages

Programming languages can vary in terms of level (high or low), generation and type. These variations in languages enable end-users to select and use specific programmes to perform certain functions in different environments.

Types of language

Different language types are shown in Figure 7.1.

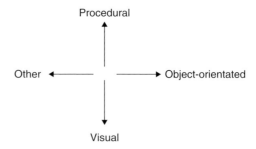

Figure 7.1 Language types

Procedural languages are the more traditional types of language, and usually the first type that is learnt. They include Fortran, Pascal, Basic and C. Procedural languages are very good for small-scale projects.

Object-orientated languages provide an excellent model for programming and designing computer software, and include C++, Smalltalk and Java.

Visual languages use images to communicate concepts based on spatial context, as opposed to a linear (text) context.

Six principles, known as Gestalt principles, should be taken into consideration if visual representations are being drawn:

- People tend to group elements together that are physically close to each other.
- People tend to group elements together that are similar in some way (e.g. same size or colour).
- People tend to see elements enclosed by lines as one unit.
- People tend to see connected elements as a single unit.
- People tend to group together elements that appear to be continuations of each other.
- People tend to make figures 'complete' when some elements are missing.

One example of a visual language is Visual Basic.Net, which is part of the Visual Studio.Net package. Screenshots showing how forms can be created using this software can be seen in Figure 7.2.

Other languages include script and markup languages that can be used for web page design, such as hypertext markup language (HTML).

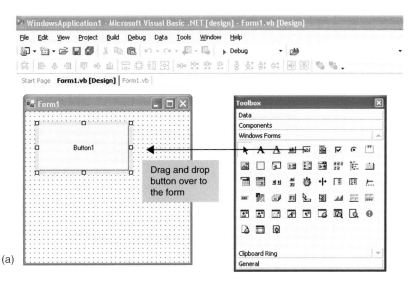

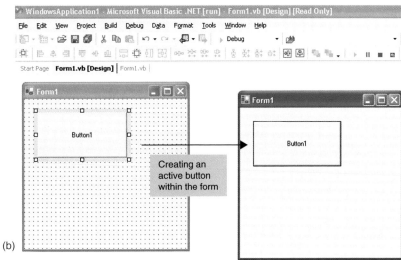

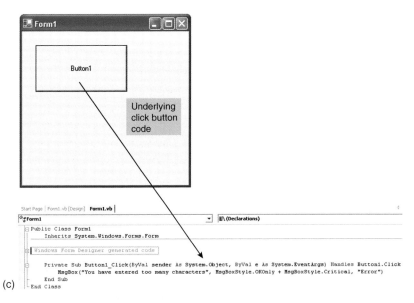

Figure 7.2 Visual Basic: (a, b) button creation; (c) the underlying code for the click button event.

Reasons for language choice

A whole host of programming languages is available, spanning different generations, levels and types. Some users have language preferences, and these can extend from:

- **the environment** – organizational, educational, scientific or social use
- **organizational policy** – where a default programming language is used as standard
- **suitability** – how functional and user friendly is it in terms of features and tools?
- **user expertise** and their level of understanding or training in a particular language
- **reliability** of the language
- **development and maintenance costs** – some languages become obsolete or unsupported
- **expandability** – is it capable of expanding with the organization or task requirement?

Features

Various features are associated with programming languages and programme design, some of which are illustrated in Table 7.1.

Table 7.1 Programming language features

Feature	Description
Variables: naming conventions, local and global, arrays	A variable is a special value or a reference that may or may not be associated with another value. Variables are named by an identifier that usually consists of alphanumeric strings. When a variable is used within a programme it can be referred to as the 'scope'
	Local variables are declared within the body of a function and can only be used within the body of that function. A global variable is located outside any programme's function
	In programming there is sometimes a requirement to handle a set of items that are of the same kind. One option is to name each item individually; however, arrays enable an entire set to be allocated the same name, with an index number distinguishing each individual item, e.g. a set of four grade scores could have the identifiers GRADE[1], GRADE[2], GRADE[3] and GRADE[4]
Loops: conditional (precheck, postcheck), fixed	A loop is a sequence of statements that can be carried out a number of times, but is only specified once. Loops can be categorized as count-controlled, condition-controlled and collection-controlled. In 'C', for example, there are three choices of loop: while, do while, for
Statements: conditional, case, assignment, input and output	A programme is constructed by a sequence of one or more statements. Statements generically can be classified as being 'simple' or 'compound'. A statement can fall into three general types: • assignment, where values (usually the results of calculations) are stored in variables • input/output, where data is read in and printed out • control, where the programme will make a decision as to what action it will take next
Operators: logical	An operator is a function that can be applied to a given value(s) to produce a result. In conjunction with 'values', operators are combined to form expressions

Data types

Data types define the format or context of the data. For example, a data type could be classed as:

- **Text** – can include any alphanumeric characters, for example abc123. Text benefits from the flexibility of combining both characters and numbers together, a good example being in an address field.
- **Integer** – a whole number such as 1, 2, 3 or 4. Benefits of using an integer are that less storage space is required, a counter can be used, mathematical operations can be performed and comparisons can be made.
- **Floating point** –a 'real number', in that it has a decimal point. The benefits of using a floating point are that percentages, areas, measurements and computations can be stored. An example of this is calculating the area of a circle: the value of pi, 3.14 … .
- **Byte** – used for storing binary data in a computer system.
- **Date** – will represent data in a specific date format that is beneficial to users as it will prompt them to enter the date in a set format, for example --/--/----, 22/02/1972.
- **Boolean** – a logic value that will return a 'true' or 'false' value. Boolean data types are very small, requiring one bit, 0 or 1, representing true or false.

A good example of using data types is when you are creating a table in a database (Figure 7.3).

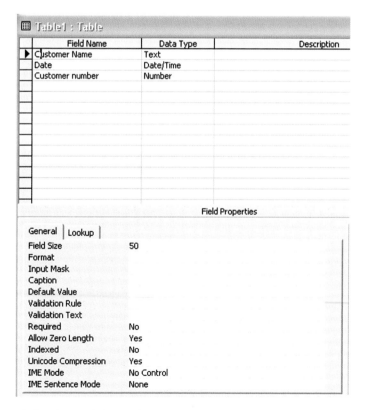

Figure 7.3 Creating a table in a database

Be able to use software design and development tools

The use of software design and development tools can ensure that a developer is following set design procedures and a particular development life cycle. Design tools can be used to provide structure and rigour to any design and ensure that the problem domain and requirements specification are being addressed.

Software development life cycle

A development process for a system, programme, software or any given project should follow a particular method or development life cycle.

A number of methods can be used when developing software; in addition, the design could follow a particular life cycle. Various life cycles exist, all of which map out various stages within a given developmental project – see chapter 7 on IT Systems Analysis and Design. For many life cycles the stages that are followed are essentially the same; however, some life cycles are more dynamic or iterative than others, which can dictate how, for example, a piece of software is designed, tested and implemented within a given environment.

Design tools
Various software design and development tools can be used to address a particular problem. Design tools such as structured diagrams, data flow diagrams (DFDs) and entity relationship modelling (ERM) are common modelling tools that can be applied to problems to generate appropriate solutions.

Structured diagrams
Structured diagrams provide a diagrammatic representation of the logic of an algorithm. The components of a structured diagram can be seen in Figure 7.4.

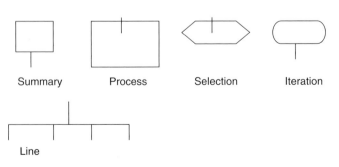

Figure 7.4 Structured diagram tools

Four tools are used in the preparation of a DFD (Figure 7.5):

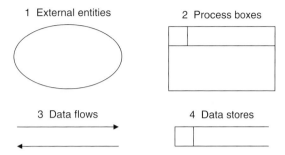

Figure 7.5 Data flow diagram tools

- **external entities** – identify people and organizations outside of the system under investigation
- **processes** – represent the activities that take place within the system
- **data flows** – provide the physical link between data sources that flow to, from and within the system
- **data stores** – detail the type of storage mechanism used to hold the data and information within the system.

External entities

External entities are used to represent people or organizations that have a role in the system but are not necessarily part of the system.

External entities can appear in a system more than once; for example, students would be part of a number of processes within a college system in terms of being enrolled on a course, having a personal tutor, being registered with student union, etc. To indicate that a student appears more than once in the system it is represented accordingly:

Processes

Process boxes represent activities that take place within the system and also activities that are linked to the system. All activities have a process attached to them: something triggers the process and an action may become the output of the process. For example, the activity of producing an assignment may be represented thus:

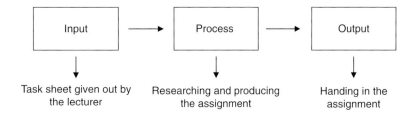

The process box has three distinct sections, each with its own identifier; for example:

- '1' identifies the process box with a unique number
- 'Checkout' provides details of the location where the activity is taking place
- 'Scan items' identifies the activity that is taking place.

Data flows

Data flows indicate the direction or flow of information within the system:

Data flows provide the link to other data flow tools within a system. These links include connecting external entities to processes and processes to data stores as identified.

Data flow links	Data store	External entity	Process
Data store	×	×	✓
External entity	×	×	✓
Process	✓	✓	×

The matrix clearly defines that the flow of information within a system must always evolve around a process. The direct flow of information from process to process is stating that two activities can take place without the intervention of an input such as a data-entry clerk.

This is true in terms of automatic processing, where a programme could receive a set of information, collate or process the information, which could then automatically trigger a second process, for example running off a report. Without the use of automated systems direct links between processes would rarely exist and therefore should not be linked on a DFD.

All data flows should be labelled clearly to identify the type of data or information that is being passed to and from sources and recipients:

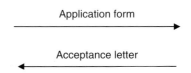

Data stores

Data stores represent different types of storage mechanisms. There are four different types of data stores:

- **D** – digitized or computerized storage mechanism, such as a file on a database
- **M** – manual storage mechanism, such as a filing cabinet
- **T(M)** – manual transient data store; a temporary manual storage mechanism such as an in-tray on a desk
- **T** – computerized data store which is temporary, e.g. e-mail which may be read once and then moved to a permanent storage file or deleted.

There are two components to a data store:

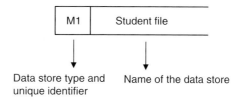

The information provided tells us that the data store type is manual, and the identifier (1) is associated with the data store mechanism, which is a customer file. If the customer file was accessed again within the system, it would then become a repeated data store. All of the information remains the same; however, we identify the repeated aspect by inserting a second line:

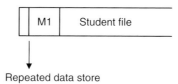

Levels of data flow diagrams

The three main levels involved with the preparation of a DFD are:

- **Level 0** (context diagram) – provides a general overview of the system and how it relates to external entities outside of the system boundary.
- **Level 1** – represents a detailed view of what and who is involved with the system. It examines what activities take place, who is involved, the data involved and appropriate storage mechanisms. Table 7.2 provides a stepped outline in developing a level 1 DFD.
- **Level 2** – provides a specific breakdown of what is happening within a specific process. The process is expanded at level 2 so that more detail can be added, giving a clear and accurate picture of a specific activity.

Step 1 Information collected from the project brief and the fact-finding investigation carried out at Store-Line Supermarkets.

Systems boundary – fresh produce

Table 7.2 Ten-step plan in preparing a level 1 data flow diagram

Step 1	Read through the information collected from: • project brief • fact-finding investigation • user catalogues
Step 2	Sort the information into clear sections identifying: • who the users are external to the system (sources and recipients of information) • what documents are used in the system • what activities take place in the system
Step 3	Produce a 'systems information table'
Step 4	Convert external users to external entities
Step 5	Convert documentation to data stores
Step 6	Convert activities to processes and identify where the activity takes place and who is involved
Step 7	Start on a small scale by looking at the input(s) and output(s) to a single process using data flows to represent the link(s) of data and information
Step 8	Position the other processes in the diagram
Step 9	Connect the remainder of the processes with their attributed input(s) and output(s)
Step 10	Check for consistency (examine initial documentation to ensure that all information has been represented and check back with users or project sponsor that the diagram is correct)

Steps 2 and 3   Systems information table:

External entities	Data stores (documents)	Processes (activities)
Head office	Promotions file	Daily meeting
Supplier	Daily stock sheet	Put out and adjust stock
	Stock adjustment sheet	Check deliveries
	Stock cabinet	Order stock
	Stock order forms	
	Stock request forms	

Steps 4, 5 and 6   Conversion of information into DFD tools:

Sample external entity   Sample data store   Sample process

Step 7   Single process DFD:

Steps 8 and 9   Connection of the remainder processes

Level 1 DFD for the fresh produce department at Store-Line Supermarket:

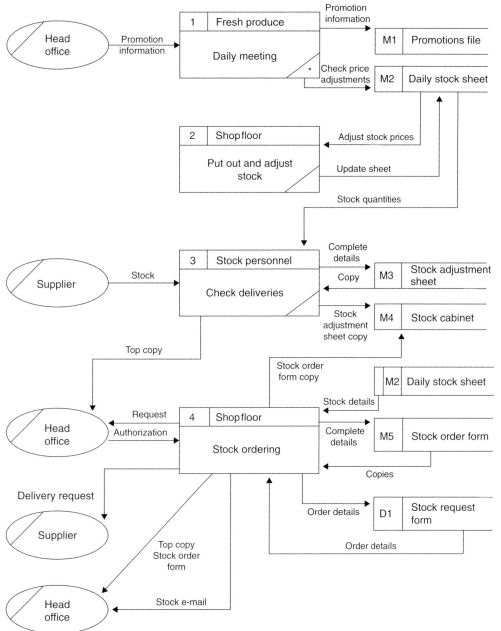

Step 10 Information is accurate and complete, checked against information provided by the personnel within the fresh produce department

The ten-step plan is a guide to preparing DFDs. Different people will use their own methods. The benefit of the plan, however, is that you are constantly re-examining the information, and understanding the system is half the battle when preparing DFDs.

Tool used for entity relationship modelling

Another tool that is used to identify problems is logical data modelling. ERM provides a detailed graphical representation of the information used within the system and identifies the relationships that exist between data items.

Similarly to data flow modelling, ERM uses a set of tools and associated textual descriptions.

The diagrammatic aspects of ERM are referred to as entity relationship diagrams. These diagrams have four main components (Figure 7.6):

- entities
- relationships
- degree
- optionality.

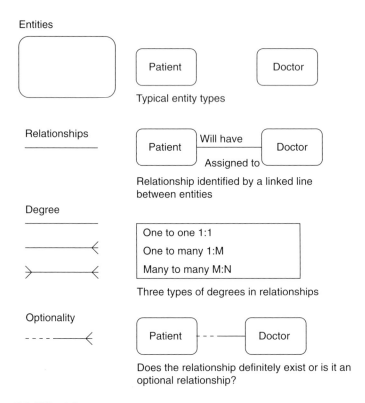

Figure 7.6 ERD notations

Entities

Entities provide the source, recipient and storage mechanism for information that is held on the system. Typical entities are shown for three systems, as follows:

Hospital system
Entities:
- patient
- doctor
- ward
- treatment
- diagnosis

College system

Entities:

- student
- registration
- course
- tutor
- enrolment

Ticket booking system

Entities:

- ticket
- seat
- booking
- performance
- reservation

Each entity will have a set of attributes that make up the information occurrences. For example:

Entity:

- student

Attributes:

- student number
- name
- address
- telephone number
- date of birth.

Each set of attributes within that entity should have a unique field that provides easy identification to the entity type. In the case of the entity type 'student' the unique key field is that of 'student number'. The unique field or 'primary key' will ensure that although two students may have the same name, no two students will have the same student number.

Relationships

To illustrate how information is used within the system, entities need to be linked together to form a relationship. The relationship between two entities could be misinterpreted; therefore labels are attached at the beginning and at the end of the relationship link to inform parties exactly what the nature of the relationship is. Entity relationships can be identified as:

Degree

There are three possible degrees of any entity relationship:

- **One to one (1:1)** – denotes that only one occurrence of each entity is used by the adjoining entity, for example a student is seen by a single tutor:

- **One to many (1:M)** – denotes that a single occurrence of one entity is linked to more than one occurrence of the adjoining entity, for example a student is seen by a number of tutors:

- **Many to many (M:N)** – denotes that many occurrences of one entity is linked to more than one occurrence of the adjoining entity, for example a student can be seen by a number of tutors, and tutors can have more than one student:

Although M:N relationships are common, the notation of linking two entities directly is adjusted and a link entity is used to connect the two, for example a customer can make a number of bookings and each of these bookings is made:

or a customer can make a number of enquiries which lead to a booking, and bookings result from a number of enquiries made by customers:

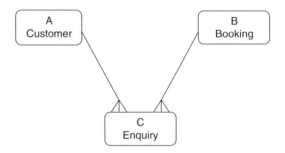

Optionality

There are two status types given to a relationship; first, those that definitely happen or exist, and secondly those that may happen or exist, this second status being referred to as 'optional'. A dashed rather than a solid link denotes optionality in a relationship:

In this scenario a customer may or may not decide to make a booking. If they do, the booking will definitely belong to/be made by a customer.

DFDs and ERM are examples of modelling tools that can be used as problem-solving tools with the software development and design stages of a given project.

Software structures

Software structures can be generic to programming languages, such as 'iteration', while other structures, such as 'objects and classes', are unique to a programming language or languages, for example C++. Software structures are important as they are the tools and formal method for writing a programme. A number of software structures can be used, including:

- iteration
- decisions
- modules
- functions
- procedures
- classes and objects.

An iteration is a repeated process for a given number of times. For example, if you wanted to display the 8 times table the same piece of code would be used each time:

> For n = 1 to 12
> print n*8
> next n

Decisions are conditions such as 'IF THEN' statements. An example of this would be if you wanted to decide whether a number was positive or negative:

> IF n < 0 THEN print 'number is positive'
> ELSE print 'number is negative'

Modules are pieces of reusable code. If, for example, a programme was designed as a module, other programmes could then use the code within that module, such as a module for maths functions or a graphics library.

Functions are pieces of code that return results, for example working out the area of a circle.

Procedures are similar to functions but they do not return a value; for example, the use of a procedure to update a bank balance for a customer.

An object is a record of different characteristics of a particular entity, such as a car (make, model, colour, registration number, mileage), house (type, age, location, number of bedrooms) or person (gender, height, weight, eye colour). A class is a combination of an object and its associated functions (for example, to return the make of the car, update the mileage or edit the service history).

Be able to design and create a programme

Designing and creating a programme can be an arduous task, especially when there are many factors that can influence the success or failure of the design. To design a successful programme, consideration needs to be given to:

- the user and environment for which the programme is being designed
- examination of the requirements specification
- techniques to be used
- constraints such as access to resources, cost or time.

If initial time is spent on planning the design and confirming the requirements with end-users then the development process should be error free.

Programme design consists of a number of elements and the relationship between these elements. If consideration has been given to each then the overall design should be a success.

Requirements specification

The requirements specification should provide a focus on a number of areas, such as:

- inputs
- outputs
- processing
- user interface
- constraints.

Inputs, outputs and processing

A system can be broken down into three component parts:

- input(s)
- processing
- output(s).

A basic systems model can be represented thus:

Inputs are resources that trigger the system. They initiate the design process and make things happen; for example:

- data and information
- people
- technology
- capital
- research and development
- a thought or idea
- raw materials or ingredients.

Processing enables the inputs to be modelled into a workable solution. Outputs reflect the end result or what is expected from the processing activity.

When designing a programme to meet the needs of a user, importance should be given to all three aspects from the input stage through to the output. A checklist (Table 7.3) could be used to address this.

Within the requirements specification the input, processing and output needs should be acknowledged and the programme should be designed around these areas.

Table 7.3 Checklist to aid an input, process, output system design

	Considered
Input(s)	
What type of data or information does the user work with?	
Is the data in a particular format?	
Does the data have to remain in this format?	
What level is the end-user in terms of being able to use any programme that is designed?	
Processing	
Will the user be familiar with the programme that is designed?	
Does the user know how to access and manipulate the new programme?	
Can the user take advantage of the functions available within the programme at a simple and complex level?	
Output(s)	
Will the programme be presented in a format that the user requires?	
Can the user output programme information easily?	
Can the user change the programme output mechanism if required?	

User interface

The user interface is of great importance when designing a programme. Consideration should be given to how the input and output screens look, the layout, colour schemes and graphics, and how well the user can interact with the programme. The user interface also needs to be fully functional to access all of the required elements of the programme.

Constraints

Programme design can be hindered by a number of constraints, such as the hardware platform in terms of compatibility and software support, lack of resources, imposed timescales and money. End-users can also constrain the development of a programme design, especially if you are working with users of differing abilities, with no clear objectives.

Design principles

When designing a programme a number of procedures should be followed, which are similar to those of a life cycle, to ensure that it is fit for purpose and fully functional. These procedures should take into consideration the following.

Step 1   Identify the programme inputs and outputs: what is the expected result for a series of data or actions? This could be prescribed by the client. Decide on or agree the criteria for the user interface design.

Step 2   Map the process/flow of data from start to end using a formal method such as flowcharting, JSP or data flow modelling.

CHAPTER 7

Activity 7.1

Read the case study about the development of the innovative design of Yahoo Music and then visit the website: http://new.uk.music.yahoo.com/

1. What do you think about the user interface? Is it functional, fast and easy to navigate? Does it draw you into the site and look aesthetically pleasing?

Process

A Community of Users

frog and Yahoo! envisioned a music application designed around discovery. frog understood that the way files are traditionally organized has little to do with the way they are actually used and experienced. And music shared from person to person and recommended by trusted community members is far more relevant and valuable than music experienced simply by walking into a store and selecting something from a rack.

The Tools for Discovery

With literally a million music files to offer the user, frog wanted to build not just literal tools for navigating content, but more flexible and innovative ways of discovering that content. Design was critical to developing new ways to explore and manage a constantly growing stream of information.

An Interface for Exploration

frog worked with Yahoo! to develop visuals, structure, interactions, and look and feel that transform the listening experience from a solitary one into one of exploration and shared community. The interface is beautiful and approachable, drawing users in to the experience.

Result

The design of Yahoo! Music design differs from other online music services by its focus on community (users can share their playlists over Yahoo's instant messenger), a powerful recommendations engine, and most significantly by its innovative use of the attributes (or metadata) of a piece of music - the genre, mood, year, community and personal ratings, etc - to guide users into un-chartered territory and delight them with new-found music. Since launch, Yahoo! Music has received industry wide acclaim, and beat out RealNetworks' Rhapsody, Napster and iTunes for the "Best Downloadable or Subscription Music Service" category in the 2005 Digital Entertainment and Media Excellence Awards.

(http://www.frogdesign.com/case-study/yahoo-music-unlimited-user-interface.html)

Step 3 Initiate the programme design based on the successful completion of steps 1 and 2. This step could involve the use of pseudocode to model the programming structures and provide a framework for the programme outline. Start to consider and design the test plan.

Step 4   Write the code for the programme (choosing data types, writing the physical code).

Step 5   Test the programme. Different programming languages lend themselves to certain design techniques, for example flowcharting would be most suitable for procedural languages such as 'C'. Therefore it is paramount that the correct design method is chosen to complement the language selected in the design process.

Activity 7.2

The administration team in the computing department of a technical college has approached you and asked whether you would like to undertake a small programming project that can support and track student grades. The team requires an application that can calculate grades automatically depending on the percentage achieved.

The grading structure is: 0–39% Fail, 40–59% Pass, 60–74% Merit and 75%+ Distinction

1. Design and create a working programme for a given scenario. The programme should show evidence of addressing each of the steps listed.
2. Justify your choice of programming language used in terms of suitability and functionality for this task.

Technical documentation

The design of a programme should be modelled around a specification of needs or requirements as set down by the project sponsor, organization or end-user. With regard to the physical design, although a development life cycle may be followed, the use of a specific design technique may be dictated by the end-user or by the culture of the organization.

To support the physical programme design technical documentation is also required to communicate the development stages, the language used, form design, code, models and underlying concepts.

Technical documentation can include:

- form design
- flowcharts
- pseudocode
- structured English
- action charts
- data dictionary
- class and instance diagrams.

Pseudocode is the programming language equivalent of 'structured English'. It provides a way of expressing a programme strategy or algorithm. The three basic algorithms that pseudocode uses are:

- **sequence** – a series of instructions that are executed sequentially with no loops or decision-making functions
- **iteration** – where a condition is tested and the task is carried out repeatedly until the result of the test is 'false'
- **selection** – after testing, a task is performed if the result of the test is 'true'.

For example:

If student's exam result is greater than or
equal to 50
Print 'pass'
else
Print 'fail'

Activity 7.3

Technical documentation can appear in a number of diagrammatic or textual formats, some of which have been listed.

Produce an information leaflet that discusses each of the following: form design, flowcharts, pseudocode, structured English, action charts, data dictionaries, class and instance diagrams.

Be able to document, test, debug and review a programmed solution

Documenting, testing, debugging and reviewing a programmed solution are crucial elements of any programme design and implementation. Testing and debugging are important to ensure that the programme works and that it meets the requirements specification. Documenting and reviewing the programme are also important to ensure that stages within the design process are transparent and easily communicated to end-users. A final review will ensure that any problems or issues have been identified and that solutions have been identified and implemented as appropriate.

Testing and debugging

Testing is paramount to the development and implementation of software. The stages of software testing will vary depending on which phase is being addressed. For example, during the prototyping phase testing may be more ad hoc as the focus is on identifying any features that are missing or defining different ways of performing a task or function. During the implementation phase, however, testing becomes more structured to identify as many faults as possible.

Throughout some software life cycles a testing model can be applied that identifies the various aspects of testing throughout analysis and design to test the requirements specification (user acceptance test) and the detailed design (unit testing). This model is known as the V-model (Figure 7.7).

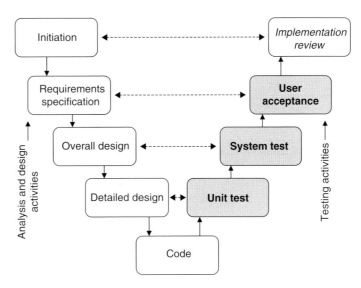

Figure 7.7 V-model of testing in relation to analysis and design

Test strategy

Before testing can begin it is normal to have a test strategy and create a test plan. The strategy may contain a series of individual tests or test cases that are specifically designed to check a particular section or module of code, or some action performed by the application.

It is not usually possible to test every possible permutation or data input/output example, so test cases are constructed to cover a range of conditions. These will include:

- **normal data** – values could be between the range of 0–10
- **extreme data** – values could be 0 or 10
- **invalid data** – where data is outside the specified range, e.g. −1 or 11.

Test plan structure

It is normal to use a table when creating and recording the results of a test programme. This table needs to identify the following:

- the programme/application name
- the tester's name
- the test date
- an identification number for each test
- a brief description of the test
- the expected outcome
- a space for the actual outcome to be recorded in
- a box to indicate Pass/Fail/Retest.

An example of this can be seen in Table 7.4.

Error messages

Error messages are an important part of any programme because they inform the user if an incorrect action or procedure has occurred. Error messages should be meaningful and clear. The integration of error messages within the programme should be comprehensive, as a number of possible error scenarios could occur as a result of user misinterpretation or incompetence.

Table 7.4 Test plan example

Application: TEY Design Screen

Tester name: Analyst 1

Test date: 31 October 2007

Test ID	Description	Expected result	Actual result	Pass/fail
	Application start/exit			
1	Application starts correctly	Display application menu screen	The screen loads correctly	P
2	All menu screen options/ buttons perform correctly	Options cause correct screens to be displayed	All options OK	P
3	Application closes down correctly	Application closes	Application fails to close properly	F

Break points – markers within the code where the programme will be instructed to stop to enable variable contents to be examined or checked for the expected values.

Specialist software tools

Debug tools are used to detect errors within a programme and track the data flow to address and correct logic or design errors. Debug tools can also be used to monitor the value of variables at certain stages within the programme using 'break points'.

User documentation and review

User documentation is important in terms of clarifying and justifying the programme design. Documentation is also required to support the end-user in understanding how the programme works. User documentation can consist of details of the hardware platform required, loading instructions, a user guide and how to access help.

Activity 7.4

Create a short user guide that supports a programme design that you have developed. The user guide should contain at least two screenshots of the programme.

In addition to user documentation, it is important to carry out a review. The review can be carried out incrementally throughout the development process or at the end of the process. A review will enable the developer to assess their design against the specification requirements, and ensure that all aspects including functionality and the user interface have been addressed.

References

The following texts should further enrich your knowledge and understanding of principles of software design and development:

Friedman, Daniel (2008) *Essentials of Programming Languages*, 3rd edn., MIT Press.

Pierce, Benjamin (2002) *Types and Programming Languages*, MIT Press.

Questions and review

1. What is the difference between a procedural language and an object-orientated language?
2. Provide three examples of why you would choose to use a specific language.
3. Give a brief description of the following:
   - Local and global variables
   - Arrays
   - Loops
   - Fixed and conditional statements
4. Explain what is meant by a data type and give three examples to support this.
5. Why is it important to follow a life cycle when you are developing a piece of software?
6. Give two examples of design tools that can be used when developing software.
7. What are the main components of a DFD?
8. Different programming languages have a range of software structures. Can you give three examples of an appropriate software structure for a specific language?
9. What are the main components of a requirements specification?
10. What appropriate technical documentation might be required when designing a program?
11. Why is it important to carry out testing and at what stage should this be done?
12. What are some of the key features of a test plan?
13. Why is it important to carry out a review at the end of the development process? What can be achieved by this?

Assessment activities

Grading criteria	Content	Suggested activity
Pass		
P1	Describe, using examples, why different types of programming languages have been developed.	Conduct research and produce a report that describes, with examples, why different programming languages have been developed.
P2	Describe, with examples, the benefits of having a variety of data types available to the programmer.	Produce an information sheet that will support the program design that describes, with examples, the benefits of having a variety of data types available to the programmer. This evidence can form part of the evidence required for P6.
P3	Write and test the functionality of a number of internally documented programs that demonstrate the features available in a given language.	Demonstrate that you can write and test the functionality of internally documented programs. The programs should demonstrate the features available in a given language. Evidence of using testing methods should be apparent in terms of a test plan or screenshots from specialised testing software.
P4	Describe the features of a programming language.	In conjunction with P1 a section could be included within the report that describes the features of a programming language.
Grading criteria	**Content**	**Suggested activity**
P5	Design and create a working programmed solution based on a defined set of requirements.	You should have evidence of designing and creating a working programmed solution based on a defined set of requirements. You should demonstrate the stages of your program design in terms of using structured methods and design techniques. The program solution must work.
P6	Document, test and review a programmed solution.	Provide evidence of documenting, testing and reviewing a programmed solution. Use documentary evidence produced for P2, P3, P5 and M2.
Merit		
M1	Compare and contrast two different types of program languages.	In conjunction with the initial report in P1, you could produce a comparative table based on two different types of program languages. The table could be accompanied by a short written justification.
M2	Justify the choice of data types used in the programmed solution.	In conjunction with P2, a short justification could be produced that looks at the data types used within the programmed solution.
M3	Adapt and improve a programmed solution based on formal testing and review.	Evidence should be provided in terms of annotated screenshots and/or an observation sheet demonstrating that the programmed solution has been formally tested and reviewed.
Distinction		
D1	Justify the language used for two different circumstances.	The final section of the report could justify the language used for two different circumstances. This could be based on the initial research carried out for P1.
D2	Evaluate a programmed solution and suggest potential further extensions.	A written evaluation should be provided to support the programmed solution. In addition potential further extensions should also be added.

Science :: Technology

Building Insights. Breaking Boundaries.

ELSEVIER

Help | Logi

| SUBJECTS | BRANDS | PRODUCTS | RESOURCES FOR | CONTAC |

Search   ▸ Advanced Search

Search   ▸ GO

All Store Categories

Full Text Search*

Full Text Search   ▸ GO

*Select Titles

Subjects

▸ Architecture & Built Environment
▸ Business
▸ Chemistry
▸ Computer Science
▸ Earth and Environmental Science
▸ Economics & Finance
▸ Engineering

What's Hot?

Websites have become one of the most popular and viewed sources of information and knowledge. Websites have the power to reach target audiences in all corners of the globe, and in conjunction with e-commerce, they also have the power to interact and provide a vehicle for trading across the world.

As websites grow in popularity, the technologies used to design and develop engaging and interactive sites also grows. In conjunction, the knowledge and technical skills required to support and maintain this provision also becomes more demanding and in some cases more specialist.

egister | Home | My Cart | U!
(0 items) | (c

Email upda
Sign up today

PRODUCTS

▸ Custom Content
▸ e-books
▸ eLearning
▸ EXCLUSIVELYe
▸ LabSuite
▸ Major Reference Works
▸ Textbooks

RESOURCES FOR

▸ Authors/Contributors
▸ Instructors
▸ Librarians
▸ Resellers
▸ Rights & Translations
▸ Societies
▸ Press

Chapter 8

Website Production and Management

The need and demand for websites have grown over recent years with the increase in organizations requiring a sales and promotion tool, the popularity of e-commerce, and the growth of general information and interactive forum sites.

In order to stay on top, the need for more sophisticated design tools is paramount as the skills and technical knowledge of developers become more advanced.

This chapter will provide an overview of the whole process from addressing a specific need to reviewing and evaluating the competed web design. The following learning outcomes will be addressed:

- Be able to design an interactive website.
- Be able to create an interactive website.
- Understand the factors that influence website performance.
- Understand the constraints related to the production and use of websites.

Be able to design an interactive website

A number of components go into the design of an interactive website. First, there needs to be an initial need or requirement: why does a website need to be designed? The rationale for this could be profit driven, information driven, market-share driven, efficiency driven or fuelled by the demands of customers. To design an interactive website you will also need a range of design tools and software that can facilitate in creating the concept, framework and interactive element.

Identification of needs

Websites are usually designed to meet a specific need or requirement. This need could be based on creating an awareness or impact, such as an information site providing details about events, current news or updates (Figure 8.1). A site could also be designed purely as a marketing tool to promote a brand, product or service linked into e-commerce facilities such as e-bay (Figure 8.2).

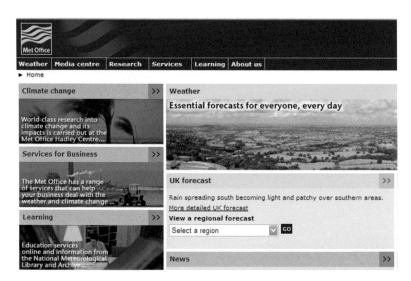

Figure 8.1 Website creating awareness: an information site (http://www.metoffice.gov.uk/)

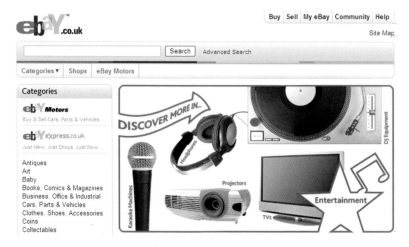

Figure 8.2 Website promoting an e-commerce linked facility (www.ebay.co.uk)

Whether the nature of the site is based on interactivity with users in terms of using online transactions and being quite dynamic, or purely a static portal for giving out information with little or no interactivity, several factors need to be considered from a design perspective.

The client and user needs may require that a website portrays a specific image, for example a corporate branding image, or an ethical or environmental image (Figure 8.3).

(a)

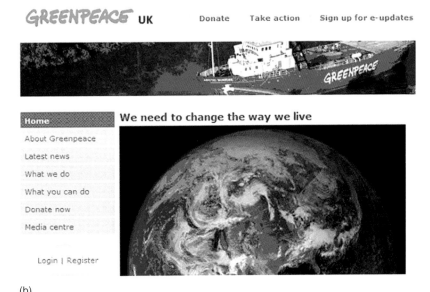

(b)

Figure 8.3　Websites that promote a particular image: (a) www.carbonfootprint.com; (b) www.greenpeace.org.uk

In terms of website development, a client or user's needs may focus on security, especially if the site is driven by the need for online transactions. Development timescales are also an issue as time equals money, so organizations may opt for a less interactive website as a tradeoff for the development time and costs being shorter.

CHAPTER 8

Once the website is operational there are issues with continued support and maintenance. Who is going to look after the site once it has been developed, and who is going to revise and update information on the site?

The presence of the website is very important; therefore, it is crucial to register with search engines and to ensure that the site is recognized easily through the choice of keywords, as this may be the only means by which users will locate the site.

Finally, the use and appropriateness of graphics are important to entice an end-user into viewing and navigating the site. Consideration should also be given to ease of navigation: how easy is it for a user to locate items, be transported to another area of the website or link to other pages, possibly external to the original site?

Activity 8.1

Websites are designed for a particular need or purpose. Carry out research into five different websites, which should include:

- commercial
- financial
- educational
- public services
- entertainment

and identify what you think their particular need or purpose is.

Design tools

When you are designing a website there are several stages and items to take into consideration. There are different ways of designing a website in terms of the tools, techniques, layouts and software that can be used.

The use of mood boards and story boards allows you to generate snapshots that can be built up to illustrate stages of a website design. Mood boards can consist of cuttings from a wide range of sources that are placed together to generate a particular feel, emotion or mood. Story boards can be designed by using graphics or by drawing images that may represent the web pages.

Activity 8.2

Use both a mood board and a story board to reflect two pages of a website. The website can be based on anything related to films and movies.

DIV and SPAN – elements that are used to logically enclose and group other elements; they are most commonly used with cascading style sheets (CSS).

Various layout techniques can be used when designing a website. The use of frames (Figure 8.4), tables, block level containers (DIV) and inline containers (SPAN) and other templates can enhance a website, provide clarity, and allow easy navigation and interaction with users and other sites.

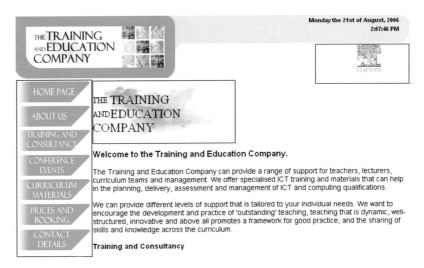

Figure 8.4 Use of frames
http://www.train-ed.co.uk

Frames are used to organize a page into sections that can easily be updated (Figure 8.4).

Colour is required in a number of areas on a web page to capture a user's attention, direct users to a specific area and promote the product or service being advertised. Colour can be applied to the background, main text, visited, non-visited and active links by using the syntax as shown:

- <body bgcolour="#code"> for background colour
- <body text="#code"> for colour of text (non-hyperlinked items)
- <body link="#code"> for colour of unvisited links
- <body vlink="#code"> for colour of visited links
- <body alink="#code"> for colour of active links (during selection).

The colour codes available to use are quite extensive, and can be selected by name or by 'hex' code.

The use of colour is very important, especially if you are trying to capture the attention of a particular audience; for example, lots of bright colours may be used in the design of a website for children (Figure 8.5). More discrete corporate colours may be used in other websites (Figure 8.6).

The screen design and content layout are also very important when designing a website. The physical positioning of text and graphics can entice people further into a site.

Software

A variety of software can be used to create interactive websites, such as using markup languages or HTML. Using HTML is very simple in that the content is typed in text and surrounded by certain commands or 'tags'. Tags will be introduced as we go through certain features of a web page.

CHAPTER 8

Figure 8.5 Colours used in a children's website
http://www.nickjr.co.uk/

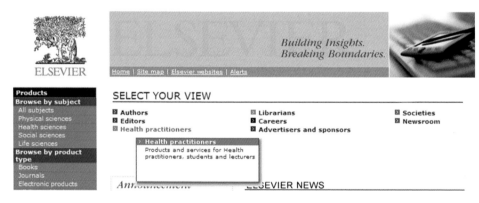

Figure 8.6 More discrete corporate colours used in professional websites

You can put a number of items on a web page, such as:

- a title
- headings
- paragraphs and line breaks
- lists and menus
- character formatting
- links
- colour.

One of the first items to put on a web page is a title. This will appear in the top of every window.

For example, if you wanted a title to be 'My Sample Web Page', you would type:

<title>My Sample Web Page</title>

In HTML, every command is surrounded by '< and >'. In most commands you also need to tell the web browser when to end this command. This can be achieved by including a slash (/) in front of the ending command, as above.

HTML is not case sensitive; therefore, <title> is the same as <TITLE>, which is the same as <tITLe>.

HTML has six levels of headings, numbered 1–6, where 1 is the largest and 6 the smallest. Headings are displayed in larger or smaller fonts, and usually bold. If you wanted to type 'My website', it would be displayed as shown in Figure 8.7.

<h1>My website</h1>

1 My website

<h2>My website</h2>

1.1 My website

<h3>My website</h3>

1.1.1 My website

<h4>My website</h4>

1.1.1.1My website

<h5>My website</h5>
1.1.1.1.1My website

<h6>My website</h6>
1.1.1.1.1.1My website

Figure 8.7 HTML headings

Paragraphs can be added by using the following syntax:

<P>This is my sample web page. I will be developing it over the next few weeks. The evidence can be used as part of my assessment</P>

will result in this:

This is my sample web page. I will be developing it over the next few weeks. The evidence can be used as part of my assessment

There may be instances where you want to end typing on one line and start on the next. This can be done by using <BR> syntax, and it is one of the few commands that does not require an ending command (/). For example:

**Welcome to my website,
What do you think?
Please leave feedback
My e-mail address is …**

The result would be

Welcome to my website,
What do you think?
Please leave feedback
My e-mail address is …

CHAPTER 8

The tag used for identifying a body of text is <body>. This is used in conjunction with paragraphs and breaks to identify clearly the content of a section of text.

```
<html>
<head>
<title>Test page</title>        Would appear in the title bar at the top
</head>
<body>
<p>My sample web page <p>
<br>
<p> What do you think?<p>
</body>
</html>
```

The sample page would look like this:

My sample web page
What do you think?

There are two types of lists that you can make in HTML: bulleted and numbered.

To make a bulleted list of: motherboard, hard disk, keyboard, memory, monitor, case, mouse, you would type:

```
<UL>
<LI> motherboard
<LI> hard disk
<LI> keyboard
<LI> memory
<LI> monitor
<LI> case
<LI> mouse.
</UL>
```

with the result being:

- motherboard
- hard disk
- keyboard
- memory
- monitor
- case
- mouse.

To make a numbered list of: motherboard, hard disk, keyboard, memory, monitor, case, mouse, you would type:

```
<OL>
<LI> motherboard
<LI> hard disk
<LI> keyboard
<LI> memory
<LI> monitor
```

\<LI\> case
\<LI\> mouse.
\</OL\>

The result being:

1. motherboard
2. hard disk
3. keyboard
4. memory
5. monitor
6. case
7. mouse.

Menus give the end-user an option of selecting items from a list without having to type information in. Two types of menus that can be used are a 'pull-down' menu and a 'scroll' menu. On a pull-down menu, if you wanted to find out the best background colour to use for a web page you could use the 'selected' command to have an initial default option and then list other colours so a user could choose a colour other than black, if preferred.

What is your preferred background for a web page?

\<SELECT NAME="**colour**"**\>**
\<OPTION\>Red
\<OPTION\>Yellow
\<OPTION\>Orange
\<OPTION\>Green
\<OPTION\>Blue
\<OPTION\>Purple
\<OPTION SELECTED\>Black   Preferred option
\<OPTION\>Brown
\</SELECT\>\<P\>

The result would be:

What is your preferred background for a web page?

However, more colour options would be available once you clicked on the arrow of the pull-down menu.

CHAPTER 8

With a scroll-down menu you can give the end-user an option of selecting more than one item. This can be done by holding down the 'command' or 'shift key' and then clicking on the items.

If a question was poséd: 'What is your favourite games console?' with the answers being XBox 360, Sony Playstation 2 or Nintendo Gamecube, the text that you would type for each list is:

What is your favourite video games console?
<SELECT NAME="video game" SIZE=3>
<OPTION VALUE="Xbox 360">Xbox 360
<OPTION VALUE="PS2">Sony Playstation 2
<OPTION VALUE="Gamecube">Nintendo Gamecube
</SELECT><P>

The result would be:

What is your favourite games console?

An example of how drop-down menus work is shown in Figures 8.8 and 8.9 with the Elsevier website (www.elsevier.com) and source code.

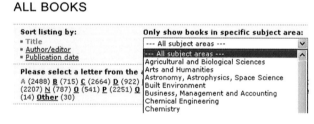

Figure 8.8 Example of a drop-down menu

At some stage you may want to change the format of your text using a range of text styles. These include:

, for *italic*
, for **bold**
<u>, </u> for <u>underlined</u>
<strike>, <strike> for ~~strikeout~~
[,] for ^{superscript}
_, for $_{subscript}$
<tt>, </tt> for teletype
<blink>, </blink> for blinking text.

Links can be identified by coloured text or graphics that take you to another location when you click over them. A link can take you to another area of the same page or to another website.

```
<tr>
<td class="verdana11DarkGrey"><b>Only show books in specific subject
area:</b></td>
</tr>

<tr>
<td><img src="/authored_framework/images/empty.gif" width="1" height="3"
border="0" alt="" /></td>
</tr>

<tr>
<td>
<select name="subject" size="1" class="locationSelect"
onchange="window.location.href=document.subjectSelect.subject[document.subject
Select.subject.selectedIndex].value">

<option
value="/wps/find/books_browse.cws_home?pseudotype=&sortBy=Title&letter=A"
selected>  ---  All subject areas  ---</option>

<option
value="/wps/find/books_browse.cws_home/L01?pseudotype=&sortBy=Title&SH1
Code=L01&letter=A">Agricultural and Biological Sciences</option>

<option
value="/wps/find/books_browse.cws_home/S01?pseudotype=&sortBy=Title&SH1
Code=S01&letter=A">Arts and Humanities</option>

<option
value="/wps/find/books_browse.cws_home/P01?pseudotype=&sortBy=Title&SH1
Code=P01&letter=A">Astronomy, Astrophysics, Space Science</option>

<option
value="/wps/find/books_browse.cws_home/P02?pseudotype=&sortBy=Title&SH1
Code=P02&letter=A">Built Environment</option>

<option
value="/wps/find/books_browse.cws_home/S02?pseudotype=&sortBy=Title&SH1
Code=S02&letter=A">Business, Management and Accounting</option>

<option
value="/wps/find/books_browse.cws_home/P03?pseudotype=&sortBy=Title&SH1
Code=P03&letter=A">Chemical Engineering</option>

<option
value="/wps/find/books_browse.cws_home/P04?pseudotype=&sortBy=Title&SH1
Code=P04&letter=A">Chemistry</option>
```

Figure 8.9 Example of drop-down menu and source code (Elsevier)

If, for example, you wanted to make a link from your web page to The Training and Education Company, the URL being:

http://www.train-ed.co.uk

you would enter:

<A HREF="http://www.train-ed.co.uk">Whatever text that you want to be coloured would be placed here</A>

The result would be:

Whatever text that you want to be coloured would be placed here

Some people have a link on their web page that will automatically send an e-mail to a certain address. To do this, if you wanted people to recognize that you had a support e-mail facility and your e-mail address was abc@123.co.uk, you could type:

<A HREF="mailto: abc@123.co.uk">Support</a>

The result being:

Support

Other software that can be used to design a website include client-side scripting languages such as JavaScript and VBScript and specialist packages such as FrontPage (Figure 8.10) or Dreamweaver.

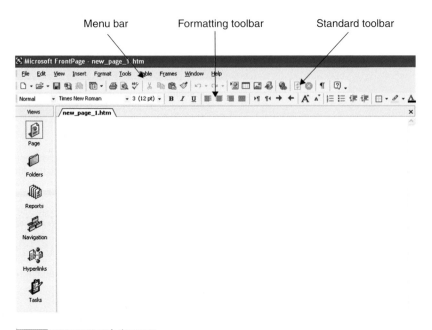

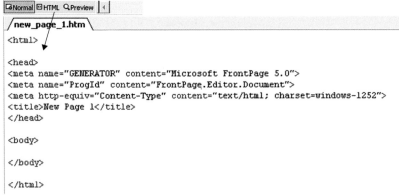

Figure 8.10 FrontPage main screen

Views available

- **Page view** – gives you a 'what you see is what you get' (WYSIWYG) editing environment for creating and editing web pages.
- **Folders view** – lists all of the files and folders in your web for easy management.
- **Reports view** – identifies any problems with the pages and links in the web pages including slow-loading pages, broken links and other errors.

CHAPTER 8

- **Navigation view** – lists the navigation order of the site and allows you to change the order in which a user will view the pages.
- **Hyperlinks view** – allows you to organize the links in the web pages.
- **Tasks view** – provides a grid for inputting tasks you need to complete on your web pages.

FrontPage is one of many specialist web design packages that can be used to support an end-user in developing a website.

Following on from the basic screen layout and views there is a range of options 'properties' that allow you to set up the layout and format certain features such as the title, colours, margins and language. You can also customize the user and system variables.

Activity 8.3

Try using HTML to design a web page that includes some formatting to promote the launch of a new 'events' and 'entertainment' site for your region.

Client-side scripting languages

Client-side scripting refers to computer programmes on the web that can be executed client-side by the user's web browser, as opposed to being on the web server, server-side. Client-side scripting languages include JavaScript and VBScript.

JavaScript supports all the structured programming syntax in C, with the exception of scoping. The main function of JavaScript is to write functions that are embedded in or included from HTML pages.

VBScript, developed by Microsoft, is classed as an 'active-scripting' language. VBScript is similar to JavaScript in terms of writing functions that are embedded in or included from HTML pages. VBScript is also used on the server-side processing of web pages.

Be able to create an interactive website

To create an interactive website you need to use a number of different tools and techniques to create a well-structured website that has a clear and dynamic layout with a range of interactive features.

Website structure

The structure of an interactive website is based on the layout of pages, navigation methods, the format of the content and cascading style sheets (CSS), and the inclusion of a range of interactive features such as a shopping cart and catalogue of products, as shown in the Amazon example in Figure 8.11, images and animations.

Figure 8.11 Example of an interactive website (http://www.amazon.co.uk)

Content

The content of any website should be clear, concise and easily accessible, as shown with the WH Smith example in Figure 8.12. As with any document the text should be proof-read and checked for accuracy and consistency throughout. In addition, any information sources used should be checked for reliability and accuracy. Permissions may also be required before data can be included on the site.

Figure 8.12 Example of a clear, concise website (http://www.whsmith.co.uk)

The structure of the content is also important in terms of the language and prose and how it is physically set out on the page; for example, is it numbered, in sections or bulleted?

Tools and techniques

The tools and techniques that can be used in interactive website design include navigation diagrams, for example linear, hierarchy or

CHAPTER 8

matrix. Navigation diagrams are used to map or provide navigation on a home page.

Building interactivity tools such as pseudocode for client server scripting is also important in the development of an interactive website. In addition, the use of animation and audiovisual elements can enhance the overall appearance and interactivity of the site.

Compliance with W3C – the World Wide Web Consortium (http://www. W3.org) – is important as W3C develops a range of technologies across a range of international communities and tries to ensure that standards about the Web and web usage are enforced.

Activity 8.4

W3C offers membership, in exchange for a number of benefits.

1. Access the W3C site www.w3.org and look at the range of services and support that is offered.
2. Do you think that membership to W3C is worthwhile? Is membership orientated more towards a certain user group, for example home users or business users?

Meta-tagging

Meta-tags provide a way to gain higher ranking within crawler-based search engines. Meta-tags offer the designer/owner of a website the ability to input certain keywords in the 'head' area of their web page. Meta-tag descriptions can also be used, with up to 250 characters being indexed.

Activity 8.5

1. Design a single-page website based on a stamp-collecting group that meets every Wednesday at the local village hall.
2. Within the head of the website page, insert an appropriate meta-tag and also include an appropriate accompanying meta-tag description.

Cascading style sheets

A style sheet is composed of a set of rules that tell a browser how to present a document. CSS is a style-sheet mechanism that has been designed to meet the specific needs of web designers and users.

Activity 8.6

CSS Zen Garden is a good example of a site that provides a number of designs that have been created using CSS.

1. Access the CSS Zen Garden site (http://www.csszengarden.com) and have a look at the various designs that have been created.

CHAPTER 8

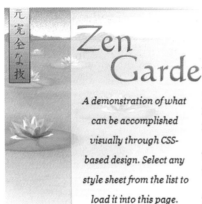

Zen Garden

The Beauty of CSS Design

A demonstration of what can be accomplished visually through CSS-based design. Select any style sheet from the list to load it into this page.

Download the sample html file and css file

The Road to Enlightenment

Littering a dark and dreary road lay the past relics of browser-specific tags, incompatible DOMs, and broken CSS support.

Today, we must clear the mind of past practices. Web enlightenment has been achieved thanks to the tireless efforts of folk like the W3C, WaSP and the major browser creators.

The css Zen Garden invites you to relax and meditate on the important lessons of the masters. Begin to see with clarity. Learn to use the (yet to be) time-honored techniques in new and invigorating fashion. Become one with the web.

2. Access and look through the resource guide: http://www.mezzoblue. com/zengarden/resources/

css Zen Garden
Resource Guide

This page used to contain a list of links to various CSS-related resources. Because of many changes to basic CSS techniques and methods since it was first built in 2003, the former list has been retired. Instead, here are other resources that offer a wide variety of modern tips and inspiration.

CSS on del.icio.us

Ever-changing links to resources and examples.

CSS on MaxDesign

Tutorials and code generation tools.

CSSBeauty

Gallery site, news, job listings, forums and more. Your one-stop shop for all things CSS.

Position is Everything

The place to go for Internet Explorer bug fixes.

Web Developer's Handbook

A really big listing of CSS and general web design-related links.

Holy CSS, Zeldman!

Another really big listing of CSS and general web design-related links.

HTML Dog

Tips and tutorials.

3. Once you have more experience in terms of designing using CSS, you can modify the CSS page by downloading the CSS files. Once created, the completed page can then be sent to CSS and uploaded onto their site.

Review

Functionality testing is one way of reviewing a website, checking the user environments, and checking the links and navigation options. If the website has been designed for a specific end-user or purpose then the

CHAPTER 8

final site should meet these requirements. Two ways of reviewing the website are to document any changes that have been forming an audit trail, and to complete test plans for each stage of the design to ensure that the designs are fit for purpose and functional.

Uploading

Finally, uploading the site using tools such as file transfer protocol (FTP) and using web servers is essential to ensure that the website is live and truly interactive. FTP is the protocol for exchanging files over the Internet. It is most commonly used for downloading a file from the server and uses the Internet to upload files to the server. The web server can accept HTTP requests from web clients known as browsers and provide an HTTP response to the client, which usually consists of an HTML document.

Understand the factors that influence website performance

A number of factors can influence website performance. The type of files and graphics used can slow the performance and loading time of a site. The actual hardware and software can also influence performance, for example the performance of the modem and the amount of memory. There are also server-side factors to consider, such as the web server capacity and the bandwidth.

File types

Different file types can be used in the design of a website; these can include image files, sound files or animation files (Figure 8.13).

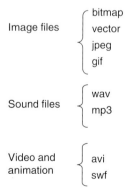

Figure 8.13 File types

Graphics can be saved in a number of formats and in some cases require conversion from one format to another to enable the image to be displayed, edited or stored. The problem with graphics files is their overall size and the amount of space that they can take up. In addition, if a graphics file is going to be sent electronically the size could inhibit file transfer, which is why graphic images are usually compressed into a suitable format.

CHAPTER 8

There is a range of different file formats, some of which are software package specific, while others are more generic.

Sound files such as MP3 and WAV (Waveform) can be used to store and transfer audio data. WAV files are quite simple as they store 'raw' audio file formats; no processing is required other than data formatting. MP3 is an audio compression technology that compresses CD quality sound by a factor of approximately 10.

Video and animation files such as AVI (audio video interleaved) and SWF (produced by Flash) can contain animation and applets, and can be used to create animated display graphics for movies and commercials.

Performance: user-side factors

User-side factors that can influence website performance include:

- modem connection speed
- PC performance factors such as cache memory and process speed.

The connection speed of a modem can influence the upload and download time of a website; the higher the connection speed, the less time it takes to carry out the process.

In terms of PC performance, the cache is faster memory that helps the main memory to keep up with the CPU. This impacts on the whole system as it stops the system slowing down and also slowing down the speed of the website.

Performance: server-side factors

From a server perspective a number of factors can impact on a website's performance. These include:

- web server capacity, for example available bandwidth
- executions to be performed before page loading.

The web server capacity can have a direct impact on a website. If the available bandwidth is too narrow then executions such as page loading will become slow.

A number of executions needs to be performed before a web page loads up. Depending on the amount and function of these executions, some pages may take longer to load than others, especially if there are lots of graphics and multimedia elements. Some pages will alert the user in terms of how long it will take to download by displaying a bar that increases in percentage value up to 100% until the page has loaded, or a message may be displayed stating that the page is loading.

Activity 8.7

The Disneyland Paris website displays a bar/graphic that alerts the user to the time required to display certain pages: http://www.disneylandparis.co.uk/index.xhtml

1. **Can you find two other websites that display a bar or an indicator as a result of executions being carried out prior to the web page display?**

Understand the constraints related to the production and use of websites

There are several constraints relating to the production of websites; some of these are security related, while others are concerned with compliance with legislation and ensuring that laws and guidelines are being adhered to. Finally, user perceptions and concerns over privacy and security of financial transactions also need to be considered as these can impose restrictions on the site construction.

Security

Security issues should be taken into consideration when designing and managing websites. From an organizational point of view there are issues concerning both internal and external threats, such as hackers and viruses. In addition, there are growing concerns about identity theft and storing information on websites, especially personal details or payment methods, as illustrated in Case study 8.1.

CHAPTER 8

Case study 8.1

Fears over website hackers

B&Q Web site lets hackers do it themselves

Munir Kotadia and Matt Loney ZDNet.co.uk
Published: 14 November 2003 17:40 GMT

A major security flaw has been exposed on home improvement retail giant B&Q's Web site, www.diy.com, which allows a potential hacker relatively easy access to its customers' personal details.

The flaw, which was discovered by a ZDNet UK reader, makes it possible to log in under accounts of other B&Q customers with little or no technical knowledge. Once logged in, it is possible to view or change the personal details of that customer, including full name, delivery address, phone number and e-mail address. Once access to an account is gained, if the customer has entered their credit card details, it is also possible to order goods on their account. B&Q customer John Dunbar said he was horrified when told about the security flaw by ZDNet UK. 'The thing is you assume that big companies like this have sorted it out, and that the security is there', he said. 'You don't for a minute think that other people can get access to your details. It's absolutely diabolical – the thought that someone could order on thousands of pounds worth of goods in my name'.

http://news.zdnet.co.uk/security/0,1000000189,39117915,00.htm

Activity 8.8

1. What security measures can be taken by an organization to reduce the threat of a security breach?
2. Most websites that deal with e-commerce provide a secure sockets layer (SSL) that protects any transaction made. Describe how SSL works.

Security protection mechanisms

Various security protection mechanisms can be used to safeguard websites. These include the use of firewalls, passwords and SSL; in addition, an awareness of and adherence to standards can help to safeguard any information that is stored or communicated.

The primary aim of a firewall is to guard against unauthorized access to an internal network. In effect, a firewall is a gateway with a lock, and the gateway only opens for information packets that pass one or more security inspections.

There are three basic types of firewall:

*   **Application gateways** – the first gateways sometimes referred to as proxy gateways. These are made up of hosts that run special software to act as a proxy server. Clients behind the firewall must know how to use the proxy, and be configured to do so in order to use Internet services. This software runs at the 'application layer' of the ISO/OSI reference model, hence the name. Traditionally, application gateways have been the most secure, because they do not allow anything to pass by default, but need to have the programmes written and turned on in order to begin passing traffic.
*   **Packet filtering** – is a technique whereby routers have 'access control lists' turned on. By default, a router will pass all traffic sent to it, and will do so without any sort of restrictions. Access control is performed at a lower ISO/OSI layer (typically, the transport or session layer). Because packet filtering is done with routers, it is often much faster than application gateways.
*   **Hybrid system** – a mixture between application gateways and packet filtering. In some of these systems, new connections must be authenticated and approved at the application layer. Once this has been done, the remainder of the connection is passed down to the session layer, where packet filters watch the connection to ensure that

CHAPTER 8

only packets that are part of an ongoing (already authenticated and approved) conversation are being passed.

SSL provides secure communications over the Internet for services such as web browsing, instant messaging, e-mail and online transactions. One of the leading SSL providers is Verisign (Figure 8.14).

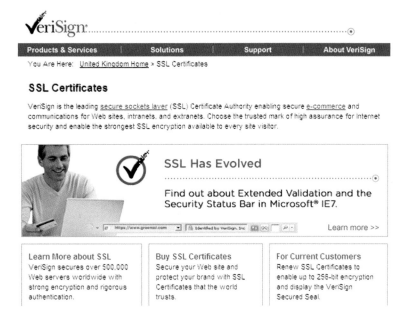

Figure 8.14 Verisign and SSL certificates
http://www.verisign.co.uk

Other measures such as passwords can be used to restrict users to certain screen or site levels, and adherence to company policies and legislation can also contribute to operating within a safe and secure environment.

Activity 8.9

Identify six websites that use SSL.

Law and guidelines

W3C compliance

The World Wide Web Consortium was founded in 1994. Its purpose is to develop 'open standards' for the Internet. Organizations can apply for membership of the consortium.

In addition to W3C, there is a wide range of legislation that can be used to protect organizations, developers and users of the Internet, including:

- Data Protection Act 1998
- Computer Misuse Act 1990
- Health and Safety Act 1984

CHAPTER 8

- Copyright Designs and Patents Act 1988
- Digital Millennium Copyright Act 2000.

Data Protection Act 1998

The Data Protection Act applies to the processing of data and information by any source, either electronic or paper based. The Act places obligations on people who collect, process and store personal records and data about consumers or customers. The Act is based on a set of principles which binds a user or an organization into following a set of procedures offering assurances that data is kept secure.

The main principles are:

1. Personal data shall be processed fairly and lawfully and, in particular, shall not be processed unless:
   - at least one of the conditions in Schedule 2 of the 1998 Act is met *and*
   - in the case of sensitive personal data, at least one of the conditions in Schedule 3 of the 1998 Act is also met.
2. Personal data shall be obtained only for one or more specified and lawful purposes, and shall not be further processed in any manner incompatible with that purpose or those purposes.
3. Personal data shall be adequate, relevant, and not excessive in relation to the purpose or purposes for which they are processed.
4. Personal data shall be accurate and, where necessary, kept up to date.
5. Personal data processed for any purpose or purposes shall not be kept for longer than is necessary for that purpose or those purposes.
6. Personal data shall be processed in accordance with the rights of data subjects under this Act.
7. Appropriate technical and organizational measures shall be taken against unauthorized or unlawful processing of personal data and against accidental loss or destruction of, or damage to, personal data.
8. Personal data shall not be transferred to a country or territory outside the EEA (European Economic Area) unless that country or territory ensures an adequate level of protection for the rights and freedoms of data subjects in relation to the processing of personal data.

The Act gives rights to individuals in respect of personal data held about them by data controllers. These include the rights:

- to make subject access requests about the nature of the information and to discover to whom it has been disclosed
- to prevent processing likely to cause damage or distress
- to prevent processing for the purposes of direct marketing
- to be informed about the mechanics of any automated decision-taking process that will significantly affect them
- not to have significant decisions that affect them made solely by an automated decision-taking process
- to take action for compensation if they suffer damage by any contravention of the Act by the data controller
- to take action to rectify, block, erase or destroy inaccurate data

> **Personal data** – information about living, identifiable individuals. Personal data does not have to be particularly sensitive information and can be as little as name and address.
> **Data users** – those who control the contents, and use of, a collection of personal data. They can be any type of company or organization, large or small, within the public or private sector. A data user can also be a sole trader, a partnership or an individual. A data user need not necessarily own a computer.
> **Data subjects** – the individuals to whom the personal data relates.
> **Automatically processed** – processed by computer or other technology such as documents image-processing systems.

- to request the Commissioner to make an assessment as to whether any provision of the Act has been contravened.

The Act provides wide exemptions for journalistic, artistic or literary purposes that would otherwise be in breach of the law.

The role of the Data Protection Commissioner

The Commissioner is an independent supervisory authority and has an international role as well as a national one. Primarily, the Commissioner is responsible for ensuring that the Data Protection legislation is enforced.

In the UK, the Commission has a range of duties, including:

- promotion of good information handling
- encouraging Codes of Practice for data controllers.

To carry out these duties the Commissioner maintains a public register of data controllers. Each register entry contains details about the controller, such as their name and address and a description of the processing of the personal data to be carried out.

Registering entries

All users, with a few exceptions, have to register an entry or entries giving their name, address and broad descriptions of:

- those about whom personal data is held
- the items of data held
- the purposes for which the data is used
- the sources from which the information may be disclosed, i.e. shown or passed to
- any overseas countries or territories to which the data may be transferred.

Computer Misuse Act 1990

The Computer Misuse Act was enacted to address the increased threat of hackers trying to gain unauthorized access to a computer system. Prior to this Act there was minimal protection and prosecution was difficult because theft of data by hacking was not considered as deprivation to the owner. Offences and penalties under this Act are listed below.

Offences
- **Unauthorized access** – an attempt by a hacker to gain unauthorized access to a computer system.
- **Unauthorized access with the intention of committing another offence** – on gaining access, a hacker will then continue with the intent of committing a further crime.
- **Unauthorized modification of data or programmes** – introducing viruses to a computer system is a criminal offence. Guilt is assessed by the level of intent to cause disruption, or to impair the processes of a computer system.

Penalties

- **Unauthorized access** – imprisonment for up to six months and/or a fine of up to £2000.
- **Unauthorized access with the intention of committing another offence** – imprisonment for up to five years and/or an unlimited fine.
- **Unauthorized modification** – imprisonment for up to five years and/or an unlimited fine.

Copyright, Designs and Patent Act 1988

The Copyright, Designs and Patent Act provides protection to software developers and organizations against unauthorized copying of their software, designs, printed material and any other product. Under copyright legislation an organization or developer can ensure that its intellectual property rights (IPR) have been safeguarded against third parties who wish to exploit and make gains from the originator's research and development.

Health and Safety Act 1984

This Act provides a framework that protects employees, and promotes a safe working environment through direct legislation and through the promotion of good practice and high standards.

Digital Millennium Copyright Act (DMCA)

The DMCA was signed into law by President Clinton on 28 October 1998. The DMCA implements two former 1996 World Intellectual Property Organization treaties: the Copyright Treaty and the Performances and Phonograms Treaty. What this means is that work posted on the Internet by the original author is protected by the Copyright Act and may also be subject to further protection under the DMCA (Case study 8.2).

Case study 8.2

DMCA case example

Russian Firm Acquitted in Landmark DMCA Case

By Keith Regan
E-Commerce Times

A Russian software firm has been acquitted of copyright infringement in the first-ever criminal prosecution under the controversial Digital Millennium Copyright

Act (DMCA). The decision seems likely to reopen debate on the law's future.

After more than two days of deliberations, a federal jury in California found ElcomSoft not guilty of copyright infringement in connection with its publishing of software that enabled users to open Adobe (Nasdaq: ADBE) eBooks for free.

Not 'willful'

The jury found that the government failed to prove that ElcomSoft willfully violated Adobe's copyright. Adobe did not immediately respond to requests for comment from the E-Commerce Times.

Although ElcomSoft sold the tool as a 'password retriever,' Adobe claimed it was a tool for circumventing password protection of eBooks, and therefore a violation of copyright protection.

The case was a lightning rod for opponents and supporters of the DMCA from its beginning in July 2001, when ElcomSoft programmer Dmitry Sklyarov was arrested outside a Las Vegas hacker convention. Sklyarov later agreed to testify against ElcomSoft in exchange for immunity.

Mixed messages

Kevin Ryan, the US attorney who prosecuted the case, said in a statement that the setback is not unexpected, given that the exact limits of the DMCA are still unclear. 'Sometimes you are going to lose, and that's what happened here', he said.

The verdict was a victory for civil rights groups and others who view the DMCA as an attempt to quell free speech.

http://www.ecommercetimes.com/story/20285.html

18 December 2002

Violation of the DMCA can result in a range of civil penalties:

- temporary and permanent injunctions
- impounding of any device or product that was involved in a violation
- statutory or actual damages
- costs
- reasonable legal fees if you are successful with any lawsuit
- destruction of any device involved in the violation of the DMCA.

On a larger scale, criminal penalties can also be imposed.

Activity 8.10

How effective do you think the DMCA is in protecting software manufacturers and developers?

User perception

Users can impose a number of constraints on the development of websites. Issues over security in terms of using a website for online transactions to buy or even sell products, and the security and privacy of information, can influence the way in which a website is designed, its content and also how the page is set out and viewed.

CHAPTER 8

CHAPTER 8

References

The following texts should further enrich your knowledge and understanding of website production and management:

Elliott, Geoff (2007) *Website Management*, Lexden Publishing.

Gray, Michael (2008) *Create Your Own Website*, Collins.

Lloyd, Ian (2006) *Build Your Own Website the Right Way Using HTML & CSS*, SITEPOINT.

Quick, Richard (2004) *Web Design in Easy Steps*, 4th edn, Computer Step.

Questions and review

1. The design of an interactive website should be driven by the identification of a need. Can you identify four possible needs?
2. What design tools can be used and what considerations should be taken into account when developing an interactive website?
3. Provide a brief overview of the following languages that can be used for website design:
   - HTML
   - Client side scripting languages
   - JavaScript
   - VBScript
4. What are cascading style sheets (CSS), and what are the benefits of using them?
5. What other structure elements would you need to consider when designing an interactive web page?
6. The content of a website is of extreme importance. What content factors and considerations would you need to address and incorporate within a design?
7. Provide an example of a storyboard – create a storyboard for a website that will be promoting various social and entertainment events in your region.
8. Why is it important for a website to have W3C compliance?
9. What does FTP mean and when is it used?
10. What factors can affect website performance?
11. Provide a brief overview of the following file types:
    - Bitmap
    - Vector
    - Jpg
    - Gif

Assessment activities

Grading criteria	Content	Suggested activity
Pass		
P1	Define the specific purpose and requirements for a website.	Produce an information booklet/guide which could be aimed at general users who are thinking about setting up and designing their own website. The first part of the leaflet could define the specific purpose and requirements for a website.
P2	Design a multi-page website to meet stated requirements.	You are required to design a multi-page website for a given user or clients need.
P3	Using a design, build a function multi-page, two-way interactive website.	The website design should be a multi-page, two-way interactive website.
P4	Review a website.	Produce a short report to accompany the website design that reviews the website which has just been designed.
P5	Describe the various factors that influence the performance of a website.	The content of the information booklet/guide can also describe the various factors that influence the performance of a website.
P6	Successfully upload a website to a web server.	This criteria requires that a website be uploaded to a web server; this can be the site that has been created in P2 and P3, or it could be a pre-created design that is uploaded.
P7	Identify the potential security issues and legal constraints involved in a particular website.	Produce a presentation that identifies the potential security issues and legal constraints involved in a particular website. Once again the website could be your own design, or it could be a site that is readily available on the internet.
Merit		
M1	Explain the tools and techniques used in the creation of a website.	In conjunction with P1 and P5, you could develop the booklet further by explaining the tools and techniques used in the creation of a website.
M2	Adapt and improve the effectiveness of a website on the basis of a formal review.	Evidence in terms of screenshots and written commentary, or an observation sheet, can be used to show that adaptations and improvements to the effectiveness of the website have been made. The review could be the stage at which the website created in P2 and P3 is examined prior to final completion.
M3	Explain techniques that can be used to minimise security risks to websites.	In conjunction with P7, additional slides could be inserted to explain techniques that could be used to minimise security risks to websites.
M4	Demonstrate that a created website meets the defined requirements and achieves the defined purpose.	In conjunction with M2, this final demonstration should illustrate that the final website meets the defined requirements as set out initially by the user/brief and achieves its defined purpose.
Distinction		
D1	Compare and evaluate two different designs created to meet a particular specification and justify the one chosen for implementation.	In conjunction with the physical website design in P2 and P3, you should demonstrate that you have followed a particular thought process in terms of selecting a specification. Provide written evidence that you have compared and evaluated two different designs for your chosen website and state how they meet the specification set out by the user/client brief. Then provide a full justification as to your selection.
D2	Produce a website that is W3C compliant.	In conjunction with P2 and P3, evidence should demonstrate that the website designed is W3C compliant.
D3	Compare 'user side' and 'server side' factors that can influence website performance.	The final section in the information booklet/guide could provide a comparative table based on the user side and server side factors that can influence website performance. This will also complement the section provided for P5.

Index

THE LIBRARY
NEW COLLEGE
SWINDON